Siddhartha Bhattacharyya, Indrajit Pan, Ashish Mani, Sourav De, Elizabeth Behrman, Susanta Chakraborti (Eds.)
Quantum Machine Learning

De Gruyter Frontiers in Computational Intelligence

Edited by
Siddhartha Bhattacharyya

Volume 6

Quantum Machine Learning

Edited by
Siddhartha Bhattacharyya, Indrajit Pan, Ashish Mani,
Sourav De, Elizabeth Behrman, Susanta Chakraborti

DE GRUYTER

Editors
Prof. (Dr.) Siddhartha Bhattacharyya
CHRIST (Deemed to be University)
Hosur Road, Bhavani Nagar, S. G. Palya,
Bangalore
560 029 Karnataka, India
dr.siddhartha.bhattacharyya@gmail.com

Dr. Indrajit Pan
RCC Institute of Information Technology
Canal South Road, Beliaghata
700 015 Kolkata, India
p.indrajit@gmail.com

Prof. (Dr.) Ashish Mani
Amity University Uttar Pradesh,
Sector-125, Noida, UP, India
amani@amity.edu

Dr. Sourav De
Cooch Behar Government Engineering College
Post Ghughumari, Harinchawara
736170 Cooch Behar, West Bengal, India
dr.sourav.de79@gmail.com

Prof. (Dr.) Elizabeth Behrman
Wichita State University
1845 Fairmount St.
67260 Wichita, USA
ecbehrman@gmail.com

Prof. (Dr.) Susanta Chakraborti
Indian Institute of Engineering, Science
and Technology
Botanic Garden
711103 Shibpur, West Bengal, India
susanta_chak@gmail.com

ISBN 978-3-11-067064-6
e-ISBN (PDF) 978-3-11-067070-7
e-ISBN (EPUB) 978-3-11-067072-1
ISSN 2512-8868

Library of Congress Control Number: 2020936493

Bibliographic information published by the Deutsche Nationalbibliothek
The Deutsche Nationalbibliothek lists this publication in the Deutsche Nationalbibliografie; detailed bibliographic data are available on the Internet at http://dnb.dnb.de.

Cover image: shulz/E+/getty images
Typesetting: Integra Software Services Pvt. Ltd.
Printing and binding: CPI books GmbH, Leck

www.degruyter.com

Siddhartha Bhattacharyya would like to dedicate this book to the memory of the deceased Jawans in Pulawama incident.

Indrajit Pan would like to dedicate this book to his research guides Prof. (Dr.) Hafizur Rahaman and Dr. Tuhina Samanta, who have always mentored him during his doctoral research.

Ashish Mani would like to dedicate this book to Most Revered Prof. Prem Saran Satsangi, Emeritus Chairman, Advisory Committee on Education, Dayalbagh Educational Institutions, India.

Sourav De would like to dedicate this book to his loving wife Debolina Ghosh and beloved son Aishik De.

Elizabeth Behrman would like to dedicate this book to EJB, JFB, and JPK; and to the memory of CFB.

Susanta Chakraborti would like to dedicate this book to his father Late Santosh Kr. Chakraborty.

Contents

List of Contributors

Siddhartha Bhattacharyya
Department of Computer Science and Engineering
CHRIST (Deemed to be University)
Bangalore, India
dr.siddhartha.bhattacharyya@gmail.com

Arit Kumar Bishwas
AIIT, Amity University, Noida
Uttar Pradesh, India
aritkumar.official@gmail.com

Amlan Chatterjee
Computer Science
California State University
Dominguez Hills, USA
achatterjee@csudh.edu

Sourav De
Department of Computer Science and Engineering
Cooch Behar Government Engineering College
Cooch Behar, West Bengal, India
dr.sourav.de79@gmail.com

Alokananda Dey
RCC Institute of Information Technology
Kolkata, West Bengal, India
alokananda_22@yahoo.co.in

Sandip Dey
Department of Computer Science
Sukanta Mahavidyalaya, Sukanta Nagar
Dhupguri, Jalpaiguri, West Bengal, India
dr.ssandip.dey@gmail.com

Bruce J. MacLennan
University of Tennessee, Knoxville
Tennessee, USA
maclennan@utk.edu

Ashish Mani
ASET, Amity University UP
Noida, Uttar Pradesh, India
amani@amity.edu

Vasile Palade
Faculty of Engineering and Computing
Coventry University
Coventry, United Kingdom
vasile.palade@coventry.ac.uk

Jan Platos
VSB Technical University of Ostrava
Czech Republic
jan.platos@vsb.cz

Vaclav Snasel
VSB Technical University of Ostrava
Czech Republic
vaclav.snasel@vsb.cz

https://doi.org/10.1515/9783110670707-203

Preface

Imparting intelligence to the machines has always been a challenging thoroughfare. Over the years, several intelligent tools have been invented or proposed to deal with the uncertainties encountered by human beings with the advent of the soft computing paradigm. However, it has been observed that even the soft computing tools often fall short in offering a reliable and reasonable solution in real time. Hence, scientists employed hybrid intelligent techniques using the combination of several soft computing tools to overcome the shortcomings.

Quantum computing has evolved from the studies of Feynman and Deutsche who evolved efficient searching techniques in the quantum domain. These searching techniques outperform the classical techniques both in terms of time and space. Inspired by this, researchers are on the spree for conjoining the existing soft computing tools with the quantum computing paradigm to evolve more robust and time efficient intelligent algorithms. The resultant algorithms are immensely useful for solving several scientific and engineering problems, which includes data processing and analysis, machine vision, social networks, big data analytics, flow shop scheduling problems to name a few.

Quantum machine learning is an emerging interdisciplinary research area which resorts to the principles of quantum physics applied to machine learning. Quantum machine learning algorithms helps to improve classical methods of machine learning by taking the advantages offered by quantum computation. Given the inherent parallelism offered due to the features of quantum computing, researchers have evolved different intelligent tools and techniques which are more robust and efficient in performance.

Quantum-enhanced machine learning refers to quantum algorithms that solve tasks in machine learning, thereby improving a classical machine learning method. Such algorithms typically require one to encode the given classical dataset into a quantum computer, so as to make it accessible for quantum information processing. After this, quantum information processing routines can be applied and the result of the quantum computation is read out by measuring the quantum system. For example, the outcome of the measurement of a qubit could reveal the result of a binary classification task. While many proposals of quantum machine learning algorithms are still purely theoretical and require a full-scale universal quantum computer to be tested, others have been implemented on small-scale or special purpose quantum devices.

This book comprises six well versed chapters from leading quantum machine learning researchers.

Chapter 1 provides an overview of the basic concepts and principles pertaining to quantum machine learning. Apart from throwing light on different aspects of quantum algorithms, the chapter also provides a bird's eye view on the principles of quantum reinforcement learning and quantum annealing. The evolution of quantum neural

https://doi.org/10.1515/9783110670707-204

networks with special mention to the pioneering works in this direction is also touched upon to enlighten the readers.

One of the most common information representations in the brain is the topographic or computational map, in which neurons are arranged systematically according to the values they represent. By representing quantitative relationships spatially, computational maps enable the brain to compute complex, nonlinear functions to the accuracy required. Chapter 2 proposes two approaches to quantum computation for machine learning by means of topographic representation. It shows how to construct unitary operators, implementable on quantum computers, that implement arbitrary (including nonlinear) functions via computational maps.

Training in machine learning techniques often requires solving a difficult optimization problem, which is the most expensive step in the entire mode-building process and its applications. One of the possible solutions in near future for reducing execution time of training process in Machine learning techniques is to implement them on quantum computers instead of classical computers. Chapter 3 discusses a global optimization technique based on Adiabatic Quantum Computation (AQC) to solve minimization of loss function without any restriction on its structure and the underlying model, which is being learned. Further, it is also shown that in the proposed framework AQC based approach would be superior to circuit-based approach in solving global optimization problems.

In Chapter 4, the authors discuss the transition from classical machine learning to quantum machine learning (QML) and explore the recent progress in this domain. QML is not only associated with the development of high-performance machine learning algorithms that can run on a quantum computer with significant performance improvements but also has a very diverse meaning in other aspects. The chapter tries to touch those aspects in brief too, but the main focus is on the advancements in the field of developing machine learning algorithms that will run on a quantum computer.

Chapter 5 is intended to present two automatic clustering techniques of image datasets, based on quantum inspired framework with two different meta-heuristic algorithms, viz., Genetic Algorithm (GA) and Bat Algorithm (BA). This work provides two novel techniques to automatically identify the optimal number of clusters present in an image dataset and also provides a comparative study between the Quantum Inspired Genetic Algorithm (QIGA) and Quantum Inspired Bat Algorithm (QIBA). A comparison is also presented between this quantum inspired algorithms with their classical counterparts. During the experiment, it is observed that the quantum inspired techniques outperform over their classical counterparts. The comparison is prepared based on the mean values of the fitness, standard deviation, standard error of the computed fitness of the cluster validity index and the optimal computational time. Finally, the superiority of the algorithms is verified in terms of the p-value which was computed from the statistical superiority test (t-test) and ranking of the proposed procedures was produced by the Friedman test. During the computation, the betterment of the fitness was judge by a well-known cluster validity index, named, DB index. The experiments are carried out on four Berkeley image and two real life grey scale image datasets.

Chapter 6 draws a line of conclusion discussing the achievable from the book. The chapter also throws light on the future trends of quantum machine learning involving multilevel quantum systems.

The editors feel that this book would come in good stead to the undergraduate and postgraduate students of computer science, information science and electronics engineering for a part of their curricula. The editors would also like to take this opportunity to render their heartfelt gratitude to De Gruyter publishing house for consenting to publish this book.

Bangalore, Kolkata, New Delhi, Kansas
November, 2019

Siddhartha Bhattacharyya
Indrajit Pan
Ashish Mani
Sourav De
Elizabeth Behrman
Susanta Chakrabarti

Sandip Dey, Sourav De, and Siddhartha Bhattacharyya

1 Introduction to quantum machine learning

Abstract: Quantum Machine Learning (QML) is popularly known to be an integrative approach to learning of the Quantum Physics (QP) and Machine Learning (ML). In this chapter, an outline of the fundamental ideas and features related to quantum machine learning is laid out. The different facets of quantum algorithms are discussed in this chapter. In addition to this, the basic features of quantum reinforcement learning and quantum annealing are also provided in this chapter. Finally, the chapter deliberates about the advancement of quantum neural networks to through light in the direction of QML.

Keywords: machine learning, Grover's Search Algorithm, reinforcement learning, quantum annealing, quantum neural networks

1.1 Quantum machine learning

Machine learning has become an emerging discipline in the recent technologies. It can be used in a variety of fields, such as computational biology, computer vision, computer security and many others. Data analysis (DA) is an equally important part in the modern industry. The ML and DA analyze data by applying statistical methods and they provide computers learning capabilities on the basis of the analysis of observed data. On the basis of learning style, Machine learning algorithms (MAA) are basically classified into different groups, viz., supervised learning (SL), unsupervised learning (UL) and semi-supervised learning (SSL). The foremost shortcomings of using ML techniques are generally known to be computational time and storage, especially which involve bulky amounts of data. In addition to these, when the current deep learning algorithms are used, the training time can turn out to be even longer. In the subsequent generation, researchers got a viable alternative by utilizing the power of quantum computers method which can be smarter enough to reduce the storage and computational time as mentioned above.

Sandip Dey, Department of Computer Science, Sukanta Mahavidyalaya, Jalpaiguri, West Bengal, India
Sourav De, Department of Computer Science & Engineering, Cooch Behar Government Engineering College, Cooch Behar, West Bengal, India
Siddhartha Bhattacharyya, Department of Computer Science and Engineering, CHRIST (Deemed to be University), Bangalore, India

https://doi.org/10.1515/9783110670707-001

Many ML problems use linear algebra to figure out matrix operations by expressing the data in matrices. Quantum computing (QC) is able to make several linear algebra computations more fast, which unreservedly improve classical ML tasks [1, 2]. Numerical methods for optimization are well admired field of research, which intend to improve the computations of the said optimization procedures. Like classical optimization, quantum optimization, a branch of QC, attempts to improve the said techniques even further. Two renowned methods of this kind are Quantum Gradient Descent (QGD) [3] and Quantum Approximate Optimization Algorithm (QAOA) [4]. These methods are efficiently applied in quantum neural networks (QNN) like Quantum Boltzman Machines [5].

Of late, the quantum machine learning, a very new field of emerging interdisciplinary research area, has come out, with the objective of coalescing the theory of machine learning with the properties of quantum computing. Quantum machine learning (QML) is intended for implementing machine learning algorithms (MLLs) in quantum atmosphere (systems) [6], by applying quantum features such as quantum superposition and quantum entanglement for solving different problems with high efficacy. Basically, QML is known to be a sub-discipline of quantum research of information processing, with the objective of introducing quantum algorithms that generally learn from data with the aim of improving existing approaches in ML. Hence, the purpose is to develop quantum applications of various MLLs, having the influence of quantum computers as well as the flexibility and learning ability of MLLs.

Several quantum algorithms have been introduced for a variety of machine learning models which include neural networks (NN), graphical models, support vector machines (SVM) so on and so forth. QML explores more basic questions about the thought of learning from the quantum perspective. In some occasions, the QML is extensively defined by researchers to apply ML to quantum information.

Besides the implicit methodologies used in QML, there exists few quantum versions of classical ML algorithms. Quantum Support Vector Machines (QSVM) [1], which is basically applied for linear classification, is a popular example of this type. In addition to that, Quantum Principle Component Analysis (QPCA) [7], a popular approach for dimensionality reduction, Quantum Guassian Mixture Models [8], another renowned approach for estimating clustering and density. An emerging sub-discipline of ML is called Deep Learning (DL). These days, quantum computers are employed for DL applications, which involve significant storage and time. Some popular examples of these applications are Quantum Boltzmann Machines [5], Quantum Generative Adversarial Networks [9], Quantum Convolutional Neural Networks [10] and Quantum Variational Autoencoders [11]. Additionally, a more enlightened field inside ML is known as Reinforcement Learning (RL). RL can be described as learning as time continues by exploring the environment.

1.2 Grover search algorithm

Grover's Search Algorithm (GAA) is a very popular quantum search algorithm [12]. This algorithm finds a group of elements which satisfy a certain condition. For performing this task, a black box, usually known as an oracle, having the capability of identifying the elements, must satisfy the criteria that are required to find. Let us suppose that the group has N elements, the oracle is called $O(M)$ times for classical computations to get all elements that satisfy the above-mentioned criteria. Using quantum mechanics, this algorithm is able to attain the same result by using $O(\sqrt{M})$ calls to the oracle.

The algorithm has the capability to make several calls to the oracle simultaneously by exploiting a special feature of quantum computing, called parallel processing. Let us assume that a search list has N number of elements. GAA uses a Hilbert space of dimension N to represent them, which can be achieved with $n = logM$ *(base 2) qubits.* Each $el \in N$ with index y is denoted by an orthonormal vector, called $|y\rangle$ in the qubits's state space. The objective is to identify the index z *of* that particular element satisfying the given search criteria.

At the outset, the oracle is being employed, in which a unitary operation having features given by

$$1. \quad if \quad y = z, U_z|y\rangle = -|y\rangle \tag{1.1a}$$

$$2. \quad if \quad y \neq z, U_z|y\rangle = |y\rangle \tag{1.1b}$$

The eqs. (1.1a) and (1.1b) can also be expressed as follows:

$$U_z = I - 2|z\rangle\langle z| \tag{1.2}$$

where, I denotes the identity operator. The GAA simultaneously uses the last oracle operator and the Grover diffusion operators, as defined by

$$U_p = 2|p\rangle\langle p| - I \tag{1.3}$$

where, $|p\rangle = \frac{1}{\sqrt{N}}\sum_{y=1}^{N} y$. The eq. (1.2) and (1.3) are utilized as follows.

At initial, the qubits' state is being initialized to the $|p\rangle$ state. Thereafter, U_z and U_p are successively applied iteratively for $r(N)$ number of times. Afterwards, the system is assessed, which provides the eigenvalue (λ_z), which may conclude the said index (z).

1.3 Quantum reinforcement learning

Besides supervised and unsupervised learning, reinforcement learning (RL) is also a popular category of learning method. In contrast to SL and UL, RL employs a scalar value called reward to assess the input-output pairs and uses the policy of trial-and error to interact with the environment for learning a mapping from states to actions. Since the year of 1980, RL has gradually become a significant approach to ML [5]. So far, it has been extensively applied in artificial intelligence (AI), particularly in robotics [13–17]. It is due to the fact that it shows an excellent performance in on-line adaptation, and in addition, it has a prevailing learning ability to any complex nonlinear systems.

There may have some complicated problems while dealing with practical applications, such as the exploration strategy, slow learning speed; in particular handling complex problems. It can be observed, especially whilst the state-action space grows to be gigantic and the number of parameters that are to be learned grows up exponentially with the increase of dimension. So far, several methods have been introduced in the recent years to combat this situation.

Different learning paradigms are united for optimizing RL. Smith [18] proposed a novel model to represent and generalize in model-less RL on the basis of self-organizing map (SOM) and benchmark Q-learning. In addition, for the sake of adaptation of fuzzy inference systems with Watkins' Q-learning for problems having large/continuous state-action spaces is also presented [17, 19, 20]. Different RL methods have been improved in practice [21–24].

Two significant quantum algorithms, called the Shor algorithm [25, 26] and the Grover algorithm [27, 28], have been introduced. Rigatos and Tzafestas [29] applied the theory of quantum computing to get the benefit of parallelization of fuzzy-logic control algorithm (FCA), whose objective is to speed up the fuzzy inference. Quantum-inspired evolutionary algorithms (QIEA) have been developed to enhance the performance of the existing evolutionary algorithms (EA) [30]. Later, Hogg and Portnov [31] introduced a quantum algorithm to solve combinatorial optimization problem having over constrained satisfiability and asymmetric traveling salesman. In recent times, the quantum search method has been applied to dynamic programming [32]. Taking into account the spirit of computation, Dong et al. [33] have developed the concept of Quantum Reinforcement Learning (QRL) inspired by the basic concept of quantum computing, called state superposition principle and parallelism. QRL was developed to speed-up learning and attaining a trade-off between exploitation and exploration of RL in the course of simulated experiments.

1.4 Quantum annealing

In statistical mechanics, quantum annealing (QA) that is applied in the field of quantum-mechanical fluctuations is the quantum version of the simulated annealing (SA) [34, 35]. It is also known as Quantum Stochastic Optimization algorithm. Like the SA, quantum annealing is also successfully applied to solve the hard optimization problem. It works on the principal of the quantum adiabatic evolution in the active field of the quantum computation and it is a profitable hybridization between classical and quantum technology. QA is applied in different fields as computer science [36], machine learning [37, 38], graph theory [39], communications [40–42], finance [43], aeronautics [44] and any others real world problems.

In simulated annealing [45], a temperature dependent random walk of a statistical-mechanical system is employed as the cost function of a given optimization problem. This function determines the potential energy profile of the solution space and thermal fluctuations avoid that the exploration gets stuck in a local minimum [46]. It is expected that the system will stay close to thermal equilibrium during time evolution. It can happen if the rate of decrease of temperature is sufficiently slow, and thus lead in the end to the zero-temperature equilibrium state, the lowest-energy state [47]. Due to the general applicability, reasonable performance, and relatively easy implementation in most cases, SA is employed effectively in many real life applications. When a problem requires an infinitely long time to derive the exact solution by keeping the system close to thermal equilibrium, SA is applied to obtain the approximate solution within a measurable computation time. The quantum fluctuations are induced in QA by introducing artificial degrees of quantum nature, non-commutative operators. The strength of the quantum fluctuations are controlled to reach the ground state by slowly reducing the temperature.

Like the finding the minimization of a cost function of a optimization problem, it can be considered as finding the ground state of a classical Ising Hamiltonian H0 [48]. Different types of practical problems have cost functions with large number of local minima. Similarly, Ising Hamiltonians are remindful of classical spin glasses [49, 50]. For these types of characteristics, it is very tough to find the global minima for classical algorithms. This problem can be overcame with the idea to elevate the classical Ising Hamiltonian H0 to the quantum domain. Based on the adiabatic theorem of quantum mechanics, the ground state of the classical Ising model can be derived by formatting the system in the ground state of some initial Hamiltonian H1 [51, 52]. It is easy to develop both theoretically and experimentally. H1 is selected in such a way it does not exchange with H0. The Hamiltonian changes gradually from H1 to H0 as the system parameters modified sufficiently slowly.

1.5 Quantum neural networks

Quantum neural networks (QNNs) are incarnations of neural network models entailing the principles of quantum mechanics. From the computational point of view [53, 54], one class of quantum neural network combine artificial neural network models and embed the features of quantum computing inside to evolve more robust and efficient models. The basic philosophy behind these efforts is to circumvent the limitations of classical neural networks in handling big data by resorting to the features of quantum parallelism, interference and entanglement. However, most of the QNN models try to replace the classical binary or the McCulloch-Pitts neurons with qubits (also called "qurons") manifested with quantum mechanical principles.

In 1995, Subhash Kak [55] and Ron Chrisley [56] put forward the idea of a quantum neural model by establishing the similarity of the neural activation function with the quantum mechanical Eigenvalue equation. Ajit Narayanan and Tammy Menneer introduced a photonic implementation of a quantum neural network using the many-universe theory which collapses into the desired states on application of quantum measurement [57]. Since then, a lot of efforts have been invested to find the quantum version of the perceptron. However, the advances in this direction was impeded due to the fact that the characteristic neural non-linear activation functions seldom follow the mathematical structure of quantum theory due to the inherent linear operations in a quantum system. After much effort, Schuld, Sinayskiy and Petruccione used the quantum phase estimation algorithm [58] to implement the activation function. Apart from this, lots of quantum-inspired models have come up to implement a fuzzy logic based neural network [59].

Elizabeth Behrman and Jim Steck [60] proposed a novel quantum computing setup comprising a number of qubits with tunable mutual interactions. In their model, the interaction strengths are updated using a training set of desired input-output relations following the classical back-propagation algorithm thereby enabling the quantum network to learn an algorithm.

The quantum associative memory was introduced by Dan Ventura and Tony Martinez in 1999 [61]. The authors proposed an algorithm to emulate an associative memory for a circuit-based quantum computer. In this algorithm, the memory states are envisaged as a superposition of quantum states. A quantum search algorithm is then used to retrieve the memory state closest to a given input. This emulation promises an exponential storage capacity of memory states.

1.6 Conclusion

This chapter provides an overview of the basic concepts and principles pertaining to quantum machine learning. Apart from throwing light on different aspects of quantum algorithms, the chapter also provides a bird's eye view on the principles of quantum reinforcement learning and quantum annealing. The evolution of quantum neural networks with special mention to the pioneering works in this direction is also touched upon to enlighten the readers.

References

[1] Harrow, A. W., Hassidim, A., & Lloyd, S. Quantum algorithm for solving linear systems of equations. Physical review letters. 2009, 15(103), 150502.

[2] Rebentrost, P., Mohseni, M., & Lloyd, S. Quantum support vector machine for big data classification. Physical review letters. 2014, 113(13), p. 130503.

[3] Kerenidis, I., & Prakash, A. Quantum gradient descent for linear systems and least squares. 2017. arXiv:1704.04992 [quant-ph].

[4] Farhi, E., Goldstone, J., & and Gutmann, S. A quantum approximate optimization algorithm. 2014. arXiv:1411.4028 [quant-ph].

[5] Amin, M. H., Andriyash, E., Rolfe, J., Kulchytskyy, B., & Melko, R. Quantum boltzmann machine. Physical review letters. 2018, 8, 021050.

[6] Schuld, M., Sinayskiy, I., & Petruccione, F. An introduction to quantum machine learning. Contemporary Physics. 2015, 56(2),172–185.

[7] Lloyd, S., Mohseni, M., & Rebentrost, P. Quantum principal component analysis. Nature Physics. 2014, 10(9),631–633.

[8] Rahman M., & and Geiger, D. Quantum clustering and gaussian mixtures. 2016. arXiv:1612.09199v1 [stat.ML].

[9] Lloyd, S., & Weedbrook, C. Quantum generative adversarial learning. Physical review letters, 2018, 040502.

[10] Cong, I., Choi, S., & Lukin, M. D. Quantum convolutional neural networks. 2018. arXiv:1810.03787v1 [quant-ph].

[11] Khoshaman, A., Vinci, W., Denis, B., Andriyash, E., & Amin, M. H. Quantum variational autoencoder. 2018. arXiv:1802.05779v1 [quant-ph].

[12] Lavor, C., Manssur, L. R. U., & Portugal, R. Grover's Algorithm: Quantum Database Search. 2003. arXiv:quant-ph/0301079.

[13] Beom, H. R., & Cho, H. S. A sensor-based navigation for a mobile robot using fuzzy logic and reinforcement learning. IEEE Transactions on Systems, Man, and Cybernetics. 1995, 25(3), 464–477.

[14] Connolly, C. I. Harmonic functions and collision probabilities. Int. J. Rob. Res. 1997, 16(4), 497–507.

[15] Smart, W. D., & Kaelbling, L. P. Effective reinforcement learning for mobile robots. Proceedings – IEEE International Conference on Robotics and Automation. 2002, 3404–3410.

[16] Kondo,T., & Ito, K. A reinforcement learning with evolutionary state recruitment strategy for autonomous mobile robots control. Robotics and Autonomous Systems. 2004, 46(2),111–124.

[17] Tzafestas, S. G., & Rigatos, G. G. Fuzzy reinforcement learning control for compliance tasks of robotic manipulators. IEEE Transactions on Systems, Man, and Cybernetics, Part B (Cybernetics). 2002, 32(1),107–113.

[18] Smith, A. J. Applications of the self-organising map to reinforcement learning. Neural Netw. 2002, 15(8/9), 1107–1124.

[19] Glorennec, P. Y, & Jouffe, L. Fuzzy Q-learning. Proceedings of 6th International Fuzzy Systems Conference. 1997, 659–662.

[20] Er, M. J, & Deng, C. Online tuning of fuzzy inference systems using dynamic fuzzy Q-learning. IEEE Transactions on Systems, Man, and Cybernetics, Part B (Cybernetics). 2004, 34(3), 1478–1489.

[21] Chen, C. L., Li, H. X., & Dong, D. Y. Hybrid control for Robot Navigation: A hierarchical Q-learning algorithm. IEEE Robotics & Automation Magazine. 2008, 15(2),37–47.

[22] Whiteson, S., & Stone, P. Evolutionary function approximation for reinforcement learning. Journal of Machine Learning Research. 2006, 7, 877–917.

[23] Kaya, M., & Alhajj, R. A novel approach to multiagent reinforcement learning: Utilizing OLAP mining in the learning process. IEEE Transactions on Systems, Man, and Cybernetics, Part C (Applications and Reviews). 2005, 35(4),582–590.
[24] Vengerov, D., Bambos, N., & Berenji, H. A fuzzy reinforcement learning approach to power control in wireless transmitters IEEE Transactions on Systems, Man, and Cybernetics, Part B (Cybernetics). 2005, 35(4),768–778.
[25] Shor, P. W. Algorithms for quantum computation: Discrete logarithms and factoring. Proceedings 35th Annual Symposium on Foundations of Computer Science. 1994, 124–134.
[26] Ekert, A., Jozsa, R. Quantum computation and Shor's factoring algorithm. Rev. Mod. Phys. 1996, 68(3),733–753.
[27] Grover, L. K. A fast quantum mechanical algorithm for database search. Proceedings of the twenty-eighth annual ACM symposium on Theory of Computing. 1996, 212–219.
[28] Grover, L. K. Quantum mechanics helps in searching for a needle in a haystack. Phys. Rev. Lett. 1997, 79(2),325–327.
[29] Rigatos, G. G., & Tzafestas, S. G. Parallelization of a fuzzy control algorithm using quantum computation. IEEE Trans. Fuzzy Syst. 2002, 10(4),451–460.
[30] Sahin, M., Atav, U., & Tomak, M. Quantum genetic algorithm method in self-consistent electronic structure calculations of a quantum dot with many electrons International Journal of Modern Physics C. 2005, 16(9),1379–1393.
[31] Hogg, T., & Portnov, D. Quantum optimization. Information Sciences. 2000, 128(3),81–197.
[32] Naguleswaran, S., White, L., & Fuss, I. Quantum search in stochastic planning. Proceedings of the Sixteenth International Conference on Automated Planning and Scheduling. 2005, 34–45.
[33] Dong, D. Y., Chen, C. L., & Chen, Z. H. Quantum reinforcement learning. Proceedings of the 1st International Conference on Natural Computing. 2005, 3611, 686–689.
[34] Matsuda, Y., Nishimori, H., & Katzgraber, H. G. Ground-state statistics from annealing algorithms: Quantum vs classical approaches. 2009. arXiv:0808.0365.
[35] Santoro, G. E., Martonak, R., Tosatti, E., & Car, R. Theory of quantum annealing of an Ising spin glass. Science. 2002, 295(5564),2427–2430.
[36] Choi, V. Adiabatic Quantum Algorithms for the NP-Complete Maximum-Weight Independent Set. Exact Cover and 3SAT Problems. 2010. http://arxiv.org/abs/1004.2226.
[37] Adachi S. H., & Henderson, M. P. Application of Quantum Annealing to Training of Deep Neural Networks. 2015. https://arxiv.org/abs/1510.06356.
[38] Amin, M. H., Andriyash, E., Rolfe, J., Kulchytskyy, B., & Melko, R. Quantum Boltzmann Machine. 2016. https://arxiv.org/abs/1601.02036.
[39] Vinci, W., Markström, K., Boixo, S., Roy, A., Spedalieri, F. M., Warburton, P. A., & Severini, S. Hearing the Shape of the Ising Model with a Programmable Superconducting-Flux Annealer. Scientific Reports. 2014, 4(5703). doi:10.1038/srep05703.
[40] Chancellor, N., Szoke, S., Vinci, W., Aeppli, G., & Warburton, P. A. Maximum-Entropy Inference with a Programmable Annealer. Scientific Reports. 2016, 6(22318). doi: 10.1038/srep22318.
[41] Otsubo, Y., Inoue, J. I., Nagata, K., & Okada, M. Effect of quantum fluctuation in error-correcting codes. Physical Review E. 2012, 86(051138). doi:10.1103/PhysRevE.86.051138.
[42] Otsubo, Y., Inoue, J. I., Nagata, K., & Okada, M. Code-division multiple-access multiuser demodulator by using quantum fluctuations. Physical Review E. 2014, 90(012126). doi:10.1103/PhysRevE.90.012126.
[43] Marzec, M. Portfolio Optimization: Applications in Quantum Computing. Social Science Research Network Technical Report. 2014. doi:10.2139/ssrn.2278729.

[44] Coxson, G. E., Hill, C. R., & Russo, J. C. Adiabatic quantum computing for finding low-peak-sidelobe codes. High Performance Extreme Computing conference, IEEE 7. 2014, 3910–3916. doi:10.1039/b509983h.
[45] Kirkpatrik, S., Gelatt Jr., C. D., & Vecchi, M. P. Optimization by simulated annealing. Science. 1983, 220, 671–680.
[46] de Falco, D., & Tamascelli, D. An introduction to quantum annealing RAIRO – Theoretical Informatics and Applications. 2011, 45, 99–116.
[47] Morita, S., & Nishimori, H. Mathematical Foundation of Quantum Annealing. 2011. https://arxiv.org/abs/0806.1859v1.
[48] Lucas, A. Ising formulations of many NP problems. Front. Physics. 2014, 12(5).
[49] Young, A. P. (Ed). Spin Glasses and Random Fields. Singapore: World Scientific. 1998.
[50] Nishimori, H. Statistical Physics of Spin Glasses and Information Processing: An Introduction. Oxford University Press, 2001.
[51] Amin, M. H. Physical Review Letters. 102, 220401, ISSN 0031-9007.
[52] Jansen, S., Ruska, M. B., & Seiler, R. Journal of Mathematical Physics. 2007, 48, 102111, ISSN 00222488.
[53] da Silva, A. J., Ludermir, T. B., & de Oliveira, W. D. Quantum perceptron over a field and neural network architecture selection in a quantum computer. Neural Networks. 2016, 76, 55–64.
[54] Panella, M., Martinelli, G. Neural networks with quantum architecture and quantum learning. International Journal of Circuit Theory and Applications. 2011, 39, 61–77.
[55] Kak, S. On quantum neural computing. Advances in Imaging and Electron Physics. 1995, 94(259).
[56] Chrisley, R. Quantum Learning. New directions in cognitive science: Proceedings of the international symposium, Saariselka, 4–9 August 1995, Lapland, Finland. Pylkkänen, P., & Pylkkö, P. (Eds). Finnish Association of Artificial Intelligence, Helsinki. 1995, 77–89.
[57] Narayanan, A., & Menneer, T. Quantum artificial neural network architectures and components. Information Sciences. 2000, 128, 231–255.
[58] Schuld, M., Sinayskiy, I., & Petruccione, F. Simulating a perceptron on a quantum computer. 2014. ArXiv:1412.3635.
[59] Purushothaman, G., & Karayiannis, N. Quantum Neural Networks (QNN's): Inherently Fuzzy Feedforward Neural Networks. IEEE Transactions on Neural Networks. 1997, 8(3),679.
[60] Behrman, E. C., Steck, J. E., Kumar, P., & Walsh, K. A. Quantum Algorithm design using dynamic learning. Quantum Information and Computation. 2008, 8(1&2),12–29.
[61] Ventura, D., & Martinez, T. A quantum associative memory based on Grover's algorithm. Proceedings of the International Conference on Artificial Neural Networks and Genetics Algorithms. 1999, 22–27.

Bruce J. MacLennan

2 Topographic representation for quantum machine learning

Abstract: One of the most common information representations in the brain is the *topographic* or *computational map*, in which neurons are arranged systematically according to the values they represent. By representing quantitative relationships spatially, computational maps enable the brain to compute complex, nonlinear functions to the accuracy required. This chapter proposes two approaches to quantum computation for machine learning by means of topographic representation. It shows how to construct unitary operators, implementable on quantum computers, that implement arbitrary (including nonlinear) functions via computational maps.

Keywords: computational map, neural network, quantum computation, quantum machine learning, quantum neural network, topographic map, topographic representation

2.1 Introduction

This chapter proposes a brain-inspired approach to quantum machine learning with the goal of circumventing many of the complications of other approaches. The fact that quantum processes are unitary presents both opportunities and challenges. A principal opportunity is that a large number of computations can be carried out in parallel in linear superposition, that is, with quantum parallelism. There are of course many technical challenges in quantum computing, but the principal theoretical challenge in quantum machine learning is the fact that quantum processes are linear whereas most approaches to machine learning depend crucially on nonlinear operations. Artificial neural networks, in particular, require a nonlinear activation function, such as a logistic sigmoid function, for nontrivial machine learning. Fortunately, the situation is not hopeless, for we know that nonlinear processes can be embedded in unitary processes, as is familiar from the circuit model of quantum computation [1].

Despite the complications of quantum machine learning, it presents a tantalizing approach to implementing large-scale machine learning in a post-Moore's law technological era. However, there are many approaches to machine learning and several

Bruce J. MacLennan, University of Tennessee, Knoxville, Tennessee, USA

https://doi.org/10.1515/9783110670707-002

approaches to quantum computation (e.g., circuit model, annealing), and it is not obvious which combinations are most likely of success. Schuld, Sinayskiy, and Petruccione [2, 3] provide useful recent reviews of some approaches to quantum neural networks and quantum machine learning more generally. They conclude that none of the proposals for quantum neural networks succeed in combining the advantages of artificial neural networks and quantum computing. They argue that open quantum systems, which involve dissipative interactions with the environment, are the most promising approach. Moreover, few of the proposals include an actual theory of quantum machine learning.

This chapter explores an approach to the quantum implementation of machine learning involving nonlinear functions operating on information represented topographically, as is common in neural cortex. However, there are many approaches to neurocomputing, that is, to brain-inspired computing, some of which may be more amenable to quantum implementation than others, and so we must say a few words about the alternatives. For the brain may be modelled at many different levels, and models at each of these levels may provide a basis for machine learning. For example, in recent years spiking neural network models have become popular once again, largely due to their power advantages when implemented in hardware [e.g., 4, 5]. Such models mimic the "all or none" action potential generation by biological neurons without addressing the detailed dynamics of the action potential. On the other hand, most contemporary applications of artificial neural networks, including those used in deep learning, use a higher level, rate-based model. That is, the real values passed between the units (neuron analogs) represent the rate of neural spiking rather than individual spikes. It has been argued that this is the appropriate level for modeling neural information processing, since there are many stochastic effects on the generation and reception of action potentials, and because the fundamental units of information processing are *microcolumns* comprising about 100 neurons [[6], chapter 2]. Therefore it is most fruitful to view neural computation as a species of massively parallel analog computation. Since quantum computation makes essential use of complex-valued probability amplitudes, it is also fruitful to treat it as a species of analog computation, and so analog information representation provides one point of contact between quantum computation and artificial neural networks [7].

2.2 Topographic representation in the brain

Another respect in which information processing in the brain differs from most artificial neural network models is that biological neural networks are spatially organized, with connectivity dependent on spatial organization. Although artificial neural networks are typically organized in layers, there is generally no spatial relationship among the neurons in each layer;[1] the exceptions are convolutional neural networks, which were in fact inspired by the organization of sensory cortex.

One of the most common spatial information organizations used by the brain is the *topographic representation* or *computational map*. In such representations, distinct points $\mathbf{x}$ in some abstract space $\mathcal{X}$ are mapped systematically to physical locations $\mathbf{r} = \mu(\mathbf{x})$ in a two-dimensional region of cortex; that is, spatial relationships among the neurons are correlated with topological relationships in the abstract space. These maps are especially common in motor areas [8] and sensory areas [[6], ch. 6]. For example, tonotopic maps have the neurons that respond to different pitches arranged in order by pitch. Retinotopic maps have a spatial organization that mirrors the organization of the retina and visual field. Neurons in primary visual cortex that respond to edges are arranged systematically according to the orientation of the edges. There are many other examples throughout the brain, and this is perhaps the single most common information representation used by the brain.

In these topographic maps, a particular value $\mathbf{x}$ is represented by an activity peak in the corresponding cortical location $\mu(\mathbf{x})$. The strength of the activity reflects the value's degree of presence, importance, or probability. Moreover, multiple simultaneous values, with differing relative strengths or probabilities, are represented by multiple simultaneous activity peaks of differing amplitudes. Therefore, such cortical maps can represent superpositions of values, each with its own amplitude.

Topographic maps provide another point of contact between artificial neural networks and quantum computation, because the computational maps in the brain are large and dense enough that they can be usefully treated mathematically as *fields*, that is, as continuous distributions of continuous quantity [9]. Such representations are suggestive of quantum mechanical wave functions, which are also continuous distributions of continuous quantity (the complex probability amplitude). In both cases these fields are treated mathematically as continuous functions on a continuous domain, and Hilbert spaces provide the mathematical framework for describing them [7]. In this chapter we exploit this analogy to implement brain-inspired approaches to quantum machine learning.

Because of their spatial representation of values, topographic maps can be used to implement arbitrary functions in the brain, essentially by a kind of table lookup. Suppose the brain needs to implement a (possibly nonlinear) transformation, $\boldsymbol{y} = f(\boldsymbol{x})$. This can be accomplished by neural connections from locations $\mathbf{r} = \mu(\mathbf{x})$ in the input map to corresponding locations $\mathbf{s} = \mu'(\mathbf{y}) = \mu'[f(\mathbf{x})]$ in the output map that represents the result space. Thus activity representing $\mathbf{x}$ in the input map will cause corresponding activity representing $\mathbf{y}$ in the output map. Moreover, a

1 They are numerically indexed, of course, but interchangeable in terms of their pattern of connections before learning.

superposition of input values will lead to a corresponding superposition of output values. Therefore, topographic representations allow the computation of nonlinear functions in linear superposition [9–13], which suggests their usefulness in quantum computation [7]. On the other hand, topographic maps make relatively inefficient use of representational resources, because every represented value has to have a location in the map [although this can be mitigated by means of *coarse coding*, by which precise values are represented by a population of broadly tuned neurons with overlapping receptive fields [14, 15]]. Therefore, use of topographic representations will require a reasonably scalable quantum computing technology. In this chapter we explore two topographic approaches to quantum computation with a focus on machine learning.

2.3 Topographic basis maps

In the brain, the state of a topographic map is a real-valued function defined over a (typically two-dimensional) space Ω. To apply these ideas in quantum computation, we consider a quantum state $|\psi\rangle$ in which the probability amplitude $\psi(\mathbf{r})$ at location $\mathbf{r} \in \Omega$ represents the amplitude (presence, importance, etc.) of the value $\mathbf{x}$ via the correspondence $\mathbf{r} = \mu(\mathbf{x})$. Here $\mathbf{r} \in \Omega$ may be a continuous index representing, for example, spatial location, or the index of a discrete quantum state, such as a wavelength or the state of a qubit register. The states $|\mathbf{r}\rangle$ form a discrete basis or continuous pseudo-basis for the input and output quantum states. In the continuous case, the input value $\mathbf{x}$ is represented by a continuous basis state $|\mu(\mathbf{x})\rangle$, which is a Dirac unit impulse at $\mathbf{r} = \mu(\mathbf{x})$, that is, $|\mathbf{r}\rangle = \delta_{\mathbf{r}}$ where $\delta_{\mathbf{r}}(\mathbf{s}) = \delta(\mathbf{s} - \mathbf{r})$. Similarly an output value $\mathbf{y}$ is represented by the continuous basis state $|\mu'(\mathbf{y})\rangle = |\mu'[f(\mathbf{x})]\rangle = \delta_{\mu'(\mathbf{y})}$. We call such a representation (whether discrete or continuous) a *topographic basis map*.

For such a continuous basis we can define a Hilbert-Schmidt linear operator:

$$T_f = \int_{\mathcal{X}} d\mathbf{x} |\mu'[f(\mathbf{x})]\rangle\langle\mu(x)|,$$

where $\mathcal{X}$ is the space of input values. (We write $T = T_f$ when f is clear from context.) This operator has the desired behavior: $T|\mu(\mathbf{x})\rangle = |\mu'[f(\mathbf{x})]\rangle$ for all $\mathbf{x} \in \mathcal{X}$. In this manner the linear operator T computes the nonlinear function f via the computational maps. We call such an operator a *graph kernel* because it uses the explicit *graph* of f [that is, the set of pairs $(\mu'[f(\mathbf{x})], \mu(\mathbf{x}))$ for all $\mathbf{x} \in \mathcal{X}$] to do a kind of table lookup.[2]

Notice that if the input map is a superposition of input values, $|\psi\rangle = a|\mu(\mathbf{x})\rangle + b|\mu(\mathbf{x}')\rangle$, then the output map will be a superposition of the corresponding results:

2 It is not the same as the graph kernels used in machine learning applied to graph theory.

$T|\psi\rangle = a|\mu'[f(\mathbf{x})]\rangle + b|\mu'[f(\mathbf{x}')]\rangle$. Therefore, continuous topographic basis maps permit nonlinear functions to be computed in linear superposition (quantum parallelism). This is a step toward quantum computation, but T might not be unitary, and we have more work to do.

The reader might question the use of a continuous basis. First, note that for separable Hilbert spaces, the continuous basis can always be replaced by an infinite discrete basis, for example, by a discrete series of sinusoids or complex exponentials of the appropriate dimension. Second, the infinite discrete basis can be approximated by a finite discrete basis, for example, by band-limited sinusoids or complex exponentials. Such an approximation is especially appropriate for neural network machine learning, which requires only low-precision calculation.

2.3.1 Bijections

We proceed to show several examples of nonlinear computations performed via quantum computational maps, beginning with a simple case and proceeding to more complex ones. For simplicity, we will ignore the representation map μ and consider computations from one quantum state to another. We consider both one-dimensional continuous domains, $\Omega = [x_l, x_u]$ and discrete domains, $\Omega = \{x_1, \ldots, x_n\}$. Typically the values would be evenly spaced, for example, $\Omega = \{0, \Delta x, 2\Delta x, \ldots, (n-1)\Delta x\}$, but this is not required, and other spacings, such as logarithmic, might be useful. (Logarithmic maps are found in some sensory regions of the cortex.) In both cases the vectors $\{|x\rangle \mid x \in \mathcal{X}\}$ are an orthonormal basis (composed of unit vectors in $\mathbb{C}^n$ for the discrete case and of delta functions in $\mathcal{L}^2(\Omega)$ for the continuous case). For example, in the discrete case, the values $x_1, \ldots, x_n$ might be represented by the composite state of an N-qubit register, where $n = 2^N$. This might seem to require a large number of qubits, but even in the absence of coarse coding, seven qubits would be sufficient to represent values with 1% precision, which is adequate for many machine learning applications.

We begin with a bijective scalar function $f:\Omega \to \Omega$, where $\Omega = [-1, 1]$. The hyperbolic tangent (appropriately restricted),[3] which is a useful sigmoid function for neural computation, is an example of such a function. The graph kernel to compute the function topographically is

$$T = \int_{\Omega} dx |f(x)\rangle\langle x|. \tag{2.1}$$

Since f is bijective, the adjoint of T is

3 For example, $f(x) = [\tanh x / \tanh 1]|_{[-1,1]}$.

$$T^{\dagger} = \int_{\Omega} dx|x\rangle\langle f(x)| = \int_{\Omega} dy|f^{-1}(y)\rangle\langle y|. \tag{2.2}$$

It is easy then to show that T is unitary and therefore amenable to quantum computation. In general, we have:

Proposition 1 The graph kernel T of a continuous bijection $f:\Omega \to \Omega'$ is unitary.

Proof: Substituting eqs. (2.1) and (2.2), observe:

$$\begin{aligned}
T^{\dagger}T &= \left(\int_{\Omega'} |f^{-1}(y)\rangle\langle y|dy\right)\left(\int_{\Omega} |f(x)\rangle\langle x|dx\right) \\
&= \int_{\Omega}\int_{\Omega'} |f^{-1}(y)\rangle\langle y|f(x)\rangle\langle x|\ dydx \\
&= \int_{\Omega}\int_{\Omega'} |f^{-1}(y)\rangle\langle y|f(x)\rangle dy\langle x|dx \\
&= \int_{\Omega} |f^{-1}(f(x))\rangle\langle x|dx \\
&= \mathrm{I}_{\Omega}.
\end{aligned}$$

The fourth line follows from the "sifting" property of the Dirac delta: $\int_{\Omega'} |F(y)\rangle\langle y|a\rangle dy = |F(a)\rangle$. Likewise,

$$\begin{aligned}
TT^{\dagger} &= \left(\int_{\Omega} |f(x)\rangle\langle x|dx\right)\left(\int_{\Omega'} |f^{-1}(y)\rangle\langle y|dy\right) \\
&= \int_{\Omega'}\int_{\Omega} |f(x)\rangle\langle x|f^{-1}(y)\rangle\langle y|\ dxdy \\
&= \int_{\Omega'} |f(f^{-1}(y))\rangle\langle y|dy = \mathrm{I}_{\Omega'}.
\end{aligned}$$

Therefore T is unitary.

In the discrete basis case, let $y_i = f(x_i)$, where f is bijective. The unit vectors $|x_i\rangle$ are a basis for $\mathbb{C}^n$, and the unit vectors $|y_i\rangle$ are also a (possibly different) basis for $\mathbb{C}^n$. The graph kernel is

$$T = \sum_{i=1}^{n} |y_i\rangle\langle x_i| = \sum_{i=1}^{n} |f(x_i)\rangle\langle x_i|, \tag{2.3}$$

which is also easily proved to be unitary. More directly, we can observe that T is a permutation matrix on the basis elements and therefore orthogonal.

If the input is a weighted superposition of values, $|\psi\rangle = \int_\Omega dx\, p(x)|x\rangle$, then applying the kernel [eq. (2.1)] will give a corresponding superposition of the outputs: $T|\psi\rangle = \int_\Omega dx\, p(x)|f(x)\rangle$. The same applies, of course, in the discrete case. Moreover, since the graph kernel is unitary, its adjoint computes the inverse function. Therefore, if $|\phi\rangle = \int_{\Omega'} dy\, q(y)|y\rangle$, then $T^\dagger|\phi\rangle = \int_{\Omega'} dy\, q(y)|f^{-1}(y)\rangle$. That is, applying the adjoint to a superposition of outputs will compute a corresponding superposition of inputs.

2.3.2 Non-surjective Injections

As a further step towards the quantum computation of arbitrary functions by means of computational maps, we consider a relatively simple case: non-surjective injections (that is, one-to-one non-onto functions). We restrict our attention to finite domains and codomains. Therefore, let $\Omega = \{x_1, \ldots, x_n\}$ and $\Omega' = \{y_1, \ldots, y_m\}$, where $n < m$, and consider an injection $f{:}\Omega \to \Omega'$. Input maps will be in an n-dimensional Hilbert space $\mathcal{H}(\Omega)$ and output maps will be in an m-dimensional Hilbert space $\mathcal{H}(\Omega')$. Since $n < m$, ancillary constants will need to be provided from a space $\mathcal{H}_C$, and so the complete input space will be in $\mathcal{H}(\Omega) \otimes \mathcal{H}_C$. Our implementation will also generate "garbage" output in a space H_G, and so the complete output space will be in $\mathcal{H}(\Omega') \otimes \mathcal{H}_G$. The input and output dimensions must be equal, and the simplest way to accomplish this is to make $\mathcal{H}_C$ m-dimensional and $\mathcal{H}_G$, n-dimensional, so that our operator is an mn-dimensional Hilbert-space transformation. Let $\{|w_1\rangle, \ldots, |w_m\rangle\}$ be an orthonormal (ON) basis for $\mathcal{H}_C$ and let $\{|v_1\rangle, \ldots, |v_n\rangle\}$ be an ON basis for $\mathcal{H}_G$. (We could in fact use the $\mathcal{H}(\Omega')$ basis for $\mathcal{H}_C$ and the $\mathcal{H}(\Omega)$ basis for $\mathcal{H}_G$, but here we develop a more general result.)

Our goal will be to define a unitary U so that $U|x\rangle|w_1\rangle = |f(x)\rangle|y\rangle$, where $|w_1\rangle$ is an ancillary constant and $|y\rangle$ is garbage. As we will see, U can be implemented by an appropriate permutation of the input basis into the output basis, which can be expressed as the sum of several operators:

$$U \stackrel{\text{def}}{=} T + S + R + Q. \tag{2.4}$$

The work of the function f is accomplished by the T component, which maps $|x_j\rangle$ to $|f(x_j)\rangle$:

$$T \stackrel{\text{def}}{=} \sum_{j=1}^{n} |f(x_j)\rangle|v_1\rangle\langle x_j|\langle w_1|. \tag{2.5}$$

Note that $T|x_j\rangle|w_1\rangle = |f(x_j)\rangle|v_1\rangle$ and T is a bijection of the n-dimensional subspace $\mathcal{H}(\Omega) \otimes \mathcal{H}(\{|w_1\rangle\})$. However, T is not unitary since it is not a surjection. See Figure 2.1.

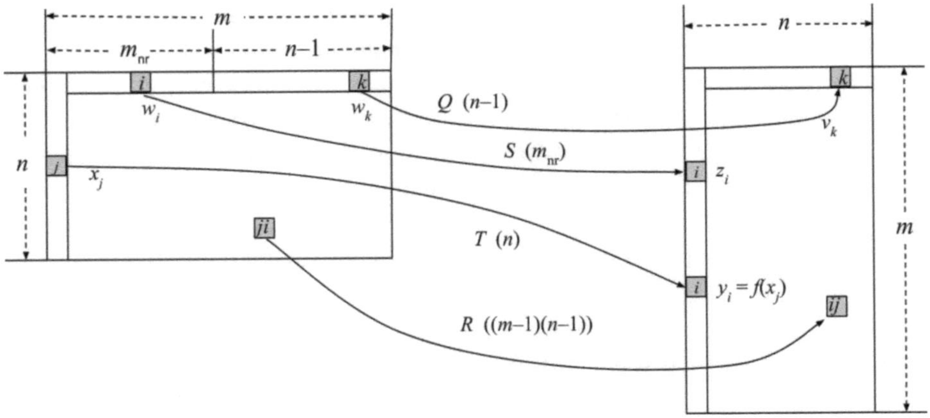

Figure 2.1: Permutation of basis vectors to implement non-surjective injections. After each component of the kernel, the number of basis vectors that it maps is indicated in parentheses; for example, *R* maps $(m-1)(n-1)$ basis vectors. *T* maps function inputs to outputs, *S* maps zero amplitudes to non-range codomain elements, and *R* and *Q* bijectively map the remaining basis elements.

The *S* component ensures that non-range elements of the codomain have preimages in the domain. Therefore, let m_{nr} be the number of codomain elements that are not in the range of *f*, that is, $m_{\text{nr}} = |\Omega' \setminus \text{Im} f| = m - n$. Call these non-range codomain elements $\{z_1, \ldots, z_{m_{\text{nr}}}\} \subset \Omega'$. Then the *S* component is defined:

$$S \stackrel{\text{def}}{=} \sum_{i=1}^{m_{\text{nr}}} |z_i\rangle |v_1\rangle \langle x_1| \langle w_{i+1}|. \tag{2.6}$$

Therefore, *S* is a bijection of an m_{nr}-dimensional subspace and each non-range element $|z_i\rangle|v_1\rangle$ has a unique preimage $|x_1\rangle|w_{i+1}\rangle$.

Note that *T* transforms *n* basis vectors (those for which the second register is $|w_1\rangle$), and *S* transforms m_{nr} basis vectors (those for which the first register is $|x_1\rangle$ and the second register is one of $|w_2\rangle, \ldots, |w_{m_{\text{nr}}+1}\rangle$), for a total of $n + m_{\text{nr}} = m$ basis elements, but the input space has a total of *mn* basis elements, and the remainder must be bijectively mapped. Therefore, to complete the unitary operator we add the following additional components:

$$R \stackrel{\text{def}}{=} \sum_{i=2}^{m} \sum_{j=2}^{n} |y_i\rangle |v_j\rangle \langle x_j| \langle w_i|, \tag{2.7}$$

$$Q \stackrel{\text{def}}{=} \sum_{j=1}^{n-1} |y_1\rangle |v_{j+1}\rangle \langle x_1| \langle w_{m_{\text{nr}}+j}|. \tag{2.8}$$

R maps $(m-1)(n-1)$ basis elements: those for $i \neq 1$ and $j \neq 1$, that is, those with neither $|x_1\rangle$ in the first register nor $|w_1\rangle$ in the second. *Q* maps the remaining $n-1$ basis elements: those which have $|x_1\rangle$ in the first register and $|w_{m_{\mathrm{nr}}+1}\rangle$ to $|w_{m_{\mathrm{nr}}+n}\rangle$ in the second (recall that $m = m_{\mathrm{nr}} + n$).

Notice that U maps every input basis vector into exactly one output basis vector and vice versa (see Figure 2.1). Summing the numbers of basis vector dyads for T, S, R, Q gives:

$$n + m_{\mathrm{nr}} + (m-1)(n-1) + (n-1) = m + (m-1)(n-1) + n - 1 = mn.$$

Proposition 2 Let Ω and Ω' be finite sets with $n = |\Omega|$, $m = |\Omega'|$, and $m > n$. Let $f{:}\Omega \to \Omega'$ be a non-surjective injection. Let $\mathcal{H}_C$ and $\mathcal{H}_G$ be Hilbert spaces of dimension m and n, respectively (representing ancillary constant inputs and garbage outputs). Let $|\omega\rangle$ be a fixed basis vector of $\mathcal{H}_C$ and $|\upsilon\rangle$ be a fixed basis vector of $\mathcal{H}_G$. Then there is a unitary operator $U \in \mathcal{L}[\mathcal{H}(\Omega) \otimes \mathcal{H}_C,\ \mathcal{H}(\Omega') \otimes \mathcal{H}_G]$ such that for all $x \in \Omega$,

$$U(|x\rangle \otimes |\omega\rangle) = |f(x)\rangle \otimes |\upsilon\rangle. \tag{2.9}$$

Proof: The proposition follows from the construction preceding the proposition.

If the input to U is a (normalized) superposition, $|\psi\rangle = \sum_k p_k |x_k\rangle$, then the output will be a superposition of the corresponding function results: $U|\psi\rangle|\omega\rangle = \sum_k p_k |f(x_k)\rangle|\upsilon\rangle$.

The dimension of the input and output spaces of this implementation is mn. A more resource-efficient but also more complicated implementation operates on a space of dimension LCM(m, n). The principle is the same: a permutation of the basis vectors.

2.3.3 Non-injective surjections

Next we consider functions $f{:}\Omega \to \Omega'$ that are surjections but not injections; that is, f maps onto Ω' but might not be one-to-one. This includes many useful functions, such as non-injective squashing functions and Gaussians, but also binary functions such as addition and multiplication (as explained later).

A non-injective function loses information, and thus for quantum computation it must be embedded in a larger injective function, which moreover must be unitary. In particular, if f is non-injective (e.g., $f(x) = f(x')$ for some $x \neq x'$), then the corresponding graph kernel will also be non-injective: $T|x\rangle = |y\rangle = T|x'\rangle$ for $|x\rangle \neq |x'\rangle$. Therefore $T(|x\rangle - |x'\rangle) = \mathbf{0}$, which implies that $|x\rangle - |x'\rangle$ is in the null space of T, which we write $\mathcal{N}(T)$. Therefore there is a bijection between the orthogonal complement of the null space, $\mathcal{N}(T)^\perp$, and the range of the operator, Im T. Hence we will implement the non-injective operation by decomposing the input $|\psi\rangle$ into orthogonal components $|\psi\rangle = |\mu\rangle + |\nu\rangle$, where $|\mu\rangle \in \mathcal{N}(T)^\perp$ and $|\nu\rangle \in \mathcal{N}(T)$. The $|\mu\rangle$

component is sufficient to determine the output, so there is a bijection $|\mu\rangle \mapsto T|\psi\rangle$, and the $|\nu\rangle$ component preserves the information to differentiate the inputs that map to this output.

To explain how this separation can be accomplished, we consider the finite-dimensional case, but it is easily extended. Let $\Omega = \{x_1, \ldots, x_n\}$ and $\Omega' = \{y_1, \ldots, y_m\}$; since f is surjective, $m \leq n$.

The desired operator T is in $\mathcal{L}(\mathcal{H}, \mathcal{H}')$, where $\mathcal{H} = \mathcal{H}(\Omega)$ is an n-dimensional Hilbert space with basis $\{|x_1\rangle, \ldots, |x_n\rangle\}$. The output space $\mathcal{H}'$ is also n-dimensional, and m of its basis vectors $\{|y_1, \ldots, |y_m\}$ are used to represent a topographic map of the function's codomain (which is also its range), $\operatorname{Im} f = \Omega' = \{y_1, \ldots, y_m\}$. Therefore $\mathcal{H}(\Omega')$ is a subspace of $\mathcal{H}'$. Let $\{|w_1\rangle, \ldots, |w_{n-m}\rangle\}$ be a basis for $\mathcal{H}(\Omega')^{\perp}$, the orthogonal complement of $\mathcal{H}(\Omega')$ in $\mathcal{H}'$. (This subspace will represent "garbage" with no computational relevance.)

We will define $\{|u_1\rangle, \ldots, |u_m\rangle\}$ to be any ON basis for $\mathcal{N}(T)^{\perp}$ (the row space of T), where m is the rank of T, and we will define $\{|v_1\rangle, \ldots, |v_{n_o}\rangle\}$ to be any ON basis for $\mathcal{N}(T)$, where $n_o = n - m$ is the nullity of T. These bases will determine the orthogonal components $|\mu\rangle \in \mathcal{N}(T)^{\perp}$ and $|\nu\rangle \in \mathcal{N}(T)$ into which any input is separated.

An example will make this clearer. Suppose $\Omega = \{k\Delta x | -N < k < N\}$ and $\Omega' = \{k\Delta x | 0 \leq k < N\}$. Let abs: $\Omega \to \Omega'$ be the absolute value function (a noninjective surjection between these sets). A basis for the nonnull space $\mathcal{N}(T)^{\perp}$ comprises $|u_0\rangle = |0\rangle$ and the vectors $|u_k\rangle = (|-k\Delta x\rangle + |k\Delta x\rangle)/\sqrt{2}$ (for $k = 1, \ldots, N-1$). (Note that $|-k\Delta x\rangle$ and $|k\Delta x\rangle$ are orthogonal vectors for $k \neq 0$.) These N basis vectors are in a one-to-one relation with the codomain elements $|k\Delta x\rangle$ (for $k = 0, \ldots, N-1$). The nullity is $n_o = (2N-1) - N = N-1$ and the basis vectors of the null space are:

$$|v_k\rangle = (|k\Delta x\rangle - |-k\Delta x\rangle)/\sqrt{2} \quad (\text{for } k = 1, \ldots, N-1).$$

Projection onto this space keeps the information necessary to distinguish the specific preimage that maps to a given output. In this case, it remembers the sign of the input: note that $\langle v_k | k\Delta x\rangle = +1/\sqrt{2}$ and $\langle v_k | -k\Delta x\rangle = -1/\sqrt{2}$. Therefore, for input $|k\Delta x\rangle$, the orthogonal components are $|\mu\rangle = |u_k\rangle/\sqrt{2}$ and $|\nu\rangle = |v_k\rangle/\sqrt{2}$; and for input $|-k\Delta x\rangle$, they are $|\mu\rangle = |u_k\rangle/\sqrt{2}$ and $|\nu\rangle = -|v_k\rangle/\sqrt{2}$. This completes the example and we return to the construction for an arbitrary non-injective surjection.

For each $y_i \in \Omega'$, let $f^{-1}\{y_i\} = \{x | f(x) = y_i\}$ be the inverse image of y_i; these are disjoint subsets of the domain Ω and correspond to orthogonal subspaces of $\mathcal{H}$. Let $n_i = |f^{-1}\{y_i\}|$ be the *preimage multiplicity* of y_i, where $n = n_1 + \cdots + n_m$. Because different $y_i \in \operatorname{Im} T$ have different preimage multiplicities, it will be convenient to separate f into m constant functions f_i: $f^{-1}\{y_i\} \to \{y_i\}$ with corresponding graph kernels:

$$T_i \stackrel{\text{def}}{=} \frac{1}{\sqrt{n_i}} \sum_{x_j \in f^{-1}\{y_i\}} |y_i\rangle\langle x_j| = |y_i\rangle \frac{1}{\sqrt{n_i}} \sum_{x_j \in f^{-1}\{y_i\}} \langle x_j| = |y_i\rangle\langle u_i|, \tag{2.10}$$

where we have defined the following normalized basis vectors of $\mathcal{N}(T)^{\perp}$:

$$|u_i\rangle \stackrel{\text{def}}{=} \frac{1}{\sqrt{n_i}} \sum_{x_j \in f^{-1}\{y_i\}} |x_j\rangle, \quad i=1, \ldots, m. \tag{2.11}$$

The constant maps T_i operate independently on orthogonal subspaces of $\mathcal{H}$.

Note that $\langle u_i | x_j \rangle \neq 0$ if and only if $y_i = f(x_j)$, and in this case $T_i|x_j\rangle = n_i^{-1/2}|y_i\rangle$. (The $n_i^{-1/2}$ factor is required for normalization of the $|u_i\rangle$.) Clearly $\{|u_1\rangle, \ldots, |u_m\rangle\}$ is an ON set, since its elements are normalized linear combinations of disjoint sets of the basis vectors. Therefore there is a one-to-one correspondence between the output vectors $|y_i\rangle$ and the basis vectors $|u_i\rangle$ of $\mathcal{N}(T)^{\perp}$.

Next we characterize $\mathcal{N}(T)$ and $\mathcal{N}(T)^{\perp}$. Observe that $|\psi\rangle \in \mathcal{N}(T_i)$ if and only if $0 = T_i|\psi\rangle = |y_i\rangle \langle u_i|\psi\rangle$, that is, if and only if $\langle u_i|\psi\rangle = 0$. Therefore, $\mathcal{N}(T_i)$ is the $n-1$ dimensional subspace orthogonal to $|u_i\rangle$ and $\mathcal{N}(T_i)^{\perp}$ is the one-dimensional subspace spanned by $\{|u_i\rangle\}$. Therefore $\mathcal{N}(T)^{\perp}$ is spanned by $\{|u_1\rangle, \ldots, |u_m\rangle\}$.

The operation T is implemented in two parts, one that handles the components representing the null space and the other that handles the components representing its orthogonal complement, the nonnull space. The first part generates "garbage," but is required for the operation to be invertible and hence unitary; the second part does the work of computing f.

We have $|v_j\rangle \in \mathcal{H}$, the basis vectors for the $\mathcal{N}(T)$ subspace of the domain, which has dimension $n_o = n - m$. Let $\{|w_j\rangle | j = 1, \ldots, n_o\}$ be any basis for $\mathcal{H}(\Omega')^{\perp}$, the orthogonal complement of $\mathcal{H}(\Omega')$ in $\mathcal{H}'$, which also has dimension $n - m$. We define $N \in \mathcal{L}(\mathcal{H}, \mathcal{H}')$ to map the null space components down to this n_o-dimensional subspace:

$$N \stackrel{\text{def}}{=} \sum_{j=1}^{n_o} |w_j\rangle \langle v_j|. \tag{2.12}$$

Since there is a one-one correspondence between basis vectors $|u_i\rangle$ and output vectors $|y_i\rangle$, we implement the function f by an operator $M \in \mathcal{L}(\mathcal{H}, \mathcal{H}')$ defined:

$$M \stackrel{\text{def}}{=} \sum_{i=1}^{m} |y_i\rangle \langle u_i|. \tag{2.13}$$

This maps the n-dimensional input into an m-dimensional output subspace. As a result the operator

$$T = M + N \in \mathcal{L}(\mathcal{H}, \mathcal{H}') \tag{2.14}$$

maps an input $|x\rangle$ to the correct output $|f(x)\rangle$, but with a scale factor and additional "garbage." Specifically, for $x_j \in f^{-1}\{y_i\}$,

$$T|x_j\rangle = (M+N)|x_j\rangle = M|x_j\rangle + N|x_j\rangle = n_i^{-1/2}|y_i\rangle + |\gamma_j\rangle. \tag{2.15}$$

where $|y_j\rangle = N|x_j\rangle$ and $\||y_j\rangle\| = \sqrt{\frac{n_i-1}{n_i}}$. Note that the garbage $|y_j\rangle$ is superimposed on the desired output. Subsequent computations operate on the $\mathcal{H}(\Omega')$ subspace and ignore the orthogonal subspace, which contains the garbage (which nevertheless must be retained, since it is entangled with the computational results).

The T operator is just a transformation from the $\{|x_1\rangle, \ldots, |x_n\rangle\}$ basis to the $\{|y_1\rangle, \ldots, |y_m\rangle, |w_1\rangle, \ldots, |w_{n-m}\rangle\}$ basis, and is obviously unitary:

$$(M+N)^\dagger(M+N) = M^\dagger M + N^\dagger M + M^\dagger N + N^\dagger N = \mathrm{I}_{\mathcal{N}(T)^\perp} + \mathbf{0} + \mathbf{0} + \mathrm{I}_{\mathcal{N}(T)} = \mathrm{I}_{\mathcal{H}},$$
$$(M+N)(M+N)^\dagger = MM^\dagger + NM^\dagger + MN^\dagger + NN^\dagger = \mathrm{I}_{\mathcal{H}(\Omega')} + \mathbf{0} + \mathbf{0} + \mathrm{I}_{\mathcal{H}(\Omega')^\perp} = \mathrm{I}_{\mathcal{H}'}.$$

Since the T operator is unitary, it can be approximated arbitrarily closely by a combination of H (Hadamard), CNOT (conditional NOT), and T ($\pi/8$) gates [[1], §4.5], or any other universal set of quantum gates.

Unfortunately, the output vectors $|y_i\rangle$ have amplitudes that depend on their preimage multiplicities. That is, if $y_i = f(x_j)$, then $T|x_j\rangle = n_i^{-1/2}|y_i\rangle + |y_j\rangle$, and we get different scale factors $n_i^{-1/2}$ depending on the preimage multiplicity. For $y_i = f(x_j)$, we define $s_i \stackrel{\text{def}}{=} n_i^{-1/2}$ to be this scale factor so that $T|x_i\rangle = s_i|y_i\rangle + |y_j\rangle$. We would like to equalize the differing amplitudes but there does not seem to be a unitary means for doing so.

It might seem that something like a Grover iteration [16] could be used to rotate the state vector from $s_i|y_i\rangle + |y_j\rangle$ to $|y_i\rangle$, but different s_i require different numbers of iterations. Something like Grover's algorithm with an unknown number of solutions could be used, but this would require trying multiple rotations. Therefore, it seems better to accept the unwanted scale factors and work with them. This means that any $|y_i\rangle$ with positive amplitudes are considered outputs from the computation, and therefore all positive amplitudes are treated the same.

If we ignore the relative magnitudes of positive amplitudes, then a quantum state $|\psi\rangle = \sum_j p_j|x_j\rangle$ (with $p_j \geq 0$) can be interpreted as representing the set of all x_j with positive amplitudes: $\{x_j \in \Omega \mid p_j > 0\}$, where we assume of course that $\sum_j p_j^2 \leq 1$. Moreover, the sum can be strictly less than one only if there are additional ancillary states that make up the difference (like $|y_j\rangle$ in the previous example). Applying T to such a state computes a state representing the image of the input set. That is, $T|\psi\rangle = \sum_j p_j s_j|f(x_j)\rangle$, which represents the set $\{f(x_j) \in \Omega | p_j > 0\}$. If $S \subseteq \Omega$ is the set represented by $|\psi\rangle$, then $T\,|\psi\rangle$ represents its image $f\,[S]$. Since zero amplitudes will always map to zero amplitudes and positive amplitudes will map to positive amplitudes, set membership will be appropriately mapped from the domain to the codomain. The preceding construction gives us:

Proposition 3 Let $\Omega = \{x_1, \ldots, x_n\}$ and $\Omega' = \{y_1, \ldots, y_m\}$ be finite sets with $n \geq m$. Suppose $\{|x_1\rangle, \ldots, |x_n\rangle\}$ is an ON basis for a Hilbert space $\mathcal{H}$ and $\{|y_1\rangle, \ldots, |y_m\rangle\}$ is an ON basis for a subspace $\mathcal{H}(\Omega')$ of $\mathcal{H}$. For any surjective function $f{:}\Omega \twoheadrightarrow \Omega'$ there is a unitary operator $T \in \mathcal{L}(\mathcal{H}, \mathcal{H})$ such that for any $x \in \Omega$,

$$T|x\rangle = \frac{1}{\sqrt{n_x}}|f(x)\rangle + |\gamma\rangle, \tag{2.16}$$

where $n_x = |f^{-1}\{f(x)\}|$, $|\gamma\rangle \in \mathcal{H}(\Omega')^\perp$, and $\||\gamma\rangle\| = \sqrt{\frac{n_x - 1}{n_x}}$.

Proof: As previously shown [eqs. (2.12)–(2.14)], this operator is given explicitly by

$$T = \sum_{i=1}^{m} |y_i\rangle\langle u_i| + \sum_{k=1}^{n-m} |w_k\rangle\langle v_k|,$$

where $|u_i\rangle \stackrel{\text{def}}{=} \frac{1}{\sqrt{n_i}} \sum_{x_j \in f^{-1}\{y_i\}} |x_j\rangle$, $n_i = |f^{-1}\{y_i\}|$, the $|v_k\rangle$ are an ON basis for the orthogonal complement of the space spanned by the $|u_i\rangle$, and the $|w_k\rangle$ are an ON basis for $\mathcal{H}(\Omega')^\perp$.

2.3.4 Arbitrary functions

In the preceding, we have assumed for convenience that the function is either non-injective or non-surjective, but not both. The solutions are easily extended to arbitrary functions, since every function can be factored as a composition of an injection and a surjection. More directly, we can combine Prop. 3 to implement a surjection onto its range with Prop. 2 to inject its range into its codomain.

Let the domain $\Omega = \{x_1, \ldots, x_n\}$, where $n = |\Omega|$, and let $\{|x_1\rangle, \ldots, |x_n\rangle\}$ be the standard basis of $\mathcal{H}(\Omega)$. Let $x_0 \notin \Omega$ be an additional value and define the extended domain $\Omega^\circ = \{x_0\} \cup \Omega$. Then $\mathcal{H}(\Omega^\circ)$ has basis $\{|x_0\rangle, |x_1\rangle, \ldots, |x_n\rangle\}$. (For example, $\mathcal{H}(\Omega^\circ)$ may be the state space of N qubits, where $n + 1 = 2^N$.) The additional x_0 input dimension will carry the null space "garbage" from previous computations. Similarly, let the codomain $\Omega' = \{y_1, \ldots, y_m\}$, where $m = |\Omega'|$, and let $\{|y_1\rangle, \ldots, |y_m\rangle\}$ be the standard basis of $\mathcal{H}(\Omega')$. Let $y_0 \notin \Omega'$ be an additional value and define the extended codomain $\Omega^* = \{y_0\} \cup \Omega'$. Then $\mathcal{H}(\Omega^*)$ has basis $\{|y_0\rangle, |y_1\rangle, \ldots, |y_m\rangle\}$. The y_0 component carries the garbage in the output state.

Let $\{r_1, \ldots, r_{m_r}\} \stackrel{\text{def}}{=} \operatorname{Im} f$ and $\{z_1, \ldots, z_{m_{nr}}\} \stackrel{\text{def}}{=} \Omega' \backslash \operatorname{Im} f$ be the range of f and its complement, respectively; since every codomain element is either a range element or not, $n = m_r + m_{nr}$. As before, an n-dimensional input $|x_j\rangle$ will be projected into orthogonal subspaces (the nonnull and null spaces) of dimension m_r and $n_o = n - m_r$, with basis vectors $|u_i\rangle$, $i = 1, \ldots, m_r$, and $|v_k\rangle$, $k = 1, \ldots, n_o$, respectively.

An additional input quantum register will be used to provide the constant zero amplitudes for non-range elements for non-surjective functions. The $m+1$ dimensional state of this ancillary register will be in, for convenience, $\mathcal{H}_C = \mathcal{H}(\Omega^*)$ with basis $\{|y_0\rangle, |y_1\rangle, \ldots, |y_m\rangle\}$. There will also be an additional output quantum register to hold the null space garbage for non-injective functions. Its $n+1$ dimensional state is in, for convenience, $\mathcal{H}_G = \mathcal{H}(\Omega^\circ)$ with basis $\{|x_0\rangle, |x_1\rangle, \ldots, |x_n\rangle\}$. Note that both the input and output spaces have dimension $(m+1)(n+1)$. This is because the ancillary input register is in the same space as the regular output register, and the ancillary output register is in the same space as the regular input register. This can be confusing because, as will be seen, we use the extra *output* vector $|y_0\rangle$ as a constant in the ancillary *input* register, and the extra *input* vector $|x_0\rangle$ appears in the ancillary *output* register.

Our goal is to define unitary $U \in \mathcal{L}[\mathcal{H}(\Omega^\circ) \otimes \mathcal{H}_C, \mathcal{H}(\Omega^*) \otimes \mathcal{H}_G]$ so that

$$U[(s|x_j\rangle + t|x_0\rangle) \otimes |y_0\rangle] = (s'|f(x_j)\rangle + t'|y_0\rangle) \otimes |\gamma\rangle, \tag{2.17}$$

for scalars s, s', t, t' and for $|\gamma\rangle \in \mathcal{H}_G$. That is, the input register is initialized to the input $|x_j\rangle$ with some positive amplitude s, possibly with superimposed garbage with amplitude t; the ancillary input register is initialized to constant $|y_0\rangle$. After computation, the output register will contain the function's value $|f(x_j)\rangle$ with some positive amplitude s'; and superimposed garbage with amplitude t'. The ancillary output register may also contain garbage. In other words, the argument to f is in the first input register [corresponding to $\mathcal{H}(\Omega^\circ)$], and its result is in the first output register [corresponding to $\mathcal{H}(\Omega^*)$], possibly with garbage in both its input $|x_0\rangle$ component and its output $|y_0\rangle$ component. The input ancillary register is initialized to a constant $|y_0\rangle$.

The work of computing f is done by the graph kernel M, which will map the $|u_i\rangle|y_0\rangle$ vectors into corresponding $(m+1)(n+1)$ dimensional result vectors $|r_i\rangle|x_0\rangle$ in the output space $\mathcal{H}(\Omega^*) \otimes \mathcal{H}_G$. (The ancillary $|x_0\rangle \in \mathcal{H}_G$ output is required so that the input and output spaces have the same dimension.) To accomplish this mapping, define M as follows:

$$M \stackrel{\text{def}}{=} \sum_{i=1}^{m_r} |r_i, x_0\rangle\langle u_i, y_0|. \tag{2.18}$$

It maps m_r of the basis vectors of $\mathcal{H}(\Omega^\circ) \otimes \mathcal{H}_C$ into m_r of the basis vectors of $\mathcal{H}(\Omega^*) \otimes \mathcal{H}_G$ (see Figure 2.2). Specifically it is a bijection between the nonnull subspace of $\mathcal{H}(\Omega^\circ) \otimes \mathcal{H}(\{|y_0\rangle\})$ and the output subspace of $\mathcal{H}(\Omega^*) \otimes \mathcal{H}(\{|x_0\rangle\})$.

Another component of the transform will map the n_0 null space basis vectors $|v_k\rangle|y_0\rangle$ of the $m+1$ dimensional $|y_0\rangle$ subspace of the input space:

$$N \stackrel{\text{def}}{=} \sum_{k=1}^{n_0} |y_0, x_k\rangle\langle v_k, y_0|. \tag{2.19}$$

It maps them to n_0 of the basis vectors $|y_0, x_k\rangle$ of the $m+1$ dimensional $|y_0\rangle$ subspace of the output space. M and N together handle non-injective functions by mapping $m_r + n_0 = n$ basis vectors.

For non-surjective functions, zero amplitudes are copied from the ancillary register $|x_0\rangle|y_i\rangle$ into the appropriate non-range codomain components $|z_i\rangle|x_0\rangle$:

$$S \stackrel{\text{def}}{=} \sum_{i=1}^{m_{nr}} |z_i, x_0\rangle\langle x_0, y_i|. \tag{2.20}$$

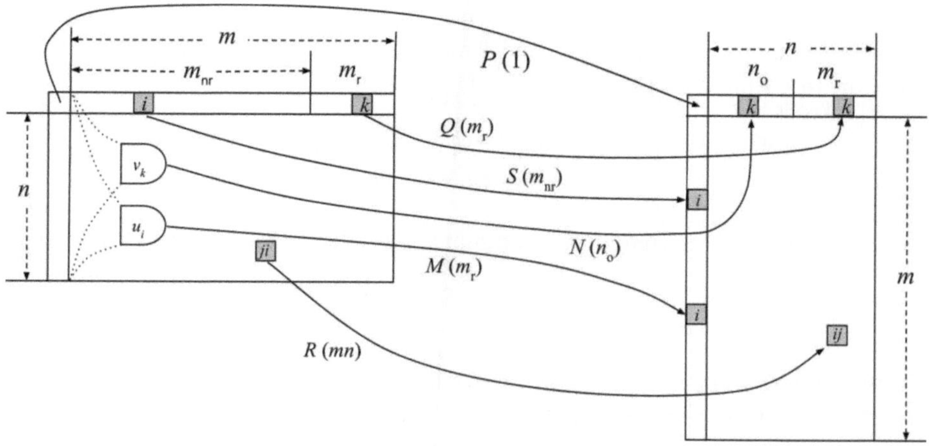

Figure 2.2: Permutation of basis vectors to implement arbitrary function. After each component of the kernel, the number of basis vectors that it maps is indicated in parentheses. For example, *R* maps *mn* basis vectors. The shapes labeled u_i and v_k represent projection onto the basis vectors of the nonnull and null spaces, respectively.

This is a mapping of m_{nr} basis vectors between the $|x_0\rangle$ subspaces of the input and output spaces. The three operators *M, N, S* handle the mapping of *f* (and disposal of the null space).

We have to be careful, however, to handle all the $|x_0\rangle|y_i\rangle$ basis vectors since m_{nr} might be less than *n* (see Figure 2.2). We map the remaining basis vectors of the $|x_0\rangle$ subspace that were not used in the *S* map to components of the $|y_0\rangle$ subspace that are unfilled by *N*:

$$Q \stackrel{\mathrm{def}}{=} \sum_{k=1}^{m_{\mathrm{r}}} |y_0, x_{n_0+k}\rangle\langle x_0, y_{m_{\mathrm{nr}}+k}|. \tag{2.21}$$

Note that $n_0 + m_{\mathrm{r}} = n = m_{\mathrm{nr}} + m_{\mathrm{r}}$ so that all these vectors are bijectively mapped.

It remains to handle the other basis vectors of the input space in a unitary way. The preceding maps have either $|x_0\rangle$ or $|y_0\rangle$, but not both, in the input register. The *R* operator maps the *mn* basis vectors with neither, $|x_j\rangle|y_i\rangle$, for $i, j \neq 0$, into their reverses:

$$R \stackrel{\mathrm{def}}{=} \sum_{i=1}^{m} \sum_{j=1}^{n} |y_i, x_j\rangle\langle x_j, y_i|. \tag{2.22}$$

The state $|x_0\rangle|y_0\rangle$ remains, and it maps to its reverse:

$$P \stackrel{\mathrm{def}}{=} |y_0, x_0\rangle\langle x_0, y_0|. \tag{2.23}$$

In summary, M bijectively maps m_{r} basis vectors, N maps n_{o} basis vectors, S maps m_{nr}, R maps mn, Q maps m_{r}, and P maps one basis vector, which accounts for all of the $(n+1)(m+1)$ basis vectors (see Figure 2.2):

$$\begin{aligned} m_{\mathrm{r}} + n_o + m_{\mathrm{nr}} + mn + m_{\mathrm{r}} + 1 &= (m_{\mathrm{r}} + n_o) + (m_{\mathrm{nr}} + m_{\mathrm{r}}) + mn + 1 \\ &= n + m + mn + 1 = (m+1)(n+1). \end{aligned}$$

The unitary operator to compute f is the sum of these linear maps [eqs. (2.18)–(2.23)]:

$$U \stackrel{\text{def}}{=} M + N + S + R + Q + P. \tag{2.24}$$

Proposition 4 Suppose $f: \Omega \to \Omega'$. Let $n = |\Omega|$ and $m = |\Omega'|$. Let $\mathcal{H}(\Omega^\circ)$ be an $n+1$ dimensional Hilbert space with basis $\{|x_0\rangle, \ldots, |x_n\rangle\}$, and let $\mathcal{H}(\Omega^*)$ be an $m+1$ dimensional space with basis $\{|y_0\rangle, \ldots, |y_m\rangle\}$. Then there a unitary operator $U \in \mathcal{L}[\mathcal{H}(\Omega^\circ) \otimes \mathcal{H}(\Omega^*),\ \mathcal{H}(\Omega^*) \otimes \mathcal{H}(\Omega^\circ)]$ so that for scalars s, t (with $|s|^2 + |t|^2 = 1$) and $x \in \Omega$:

$$U[(s|x\rangle + t|x_0\rangle) \otimes |y_0\rangle] = ss'|f(x), x_0\rangle + t'|y_0, \gamma\rangle, \tag{2.25}$$

where $s' \stackrel{\text{def}}{=} n_x^{-1/2}$, where $n_x = |f^{-1}\{f(x)\}|$, and $|ss'|^2 + |t'|^2 = 1$.

Proof: By construction [eqs. (2.18) and (2.19)] we know:

$$\begin{aligned} U|x_j, y_0\rangle &= (M+N)|x_j, y_0\rangle \\ &= \left(\sum_{i=1}^{m_r} |r_i, x_0\rangle\langle u_i, y_0|\right)|x_j, y_0\rangle + \left(\sum_{k=1}^{n_o} |y_0, x_k\rangle\langle v_k, y_0|\right)|x_j, y_0\rangle \\ &= |f(x_j), x_0\rangle\langle u_i|x_j\rangle + |y_0\rangle \sum_{k=1}^{n_o} |x_k\rangle\langle v_k|x_j\rangle \\ &= s_j|f(x_j), x_0\rangle + |y_0, \gamma\rangle, \end{aligned}$$

where $|\gamma\rangle = \left(\sum_{k=1}^{n_o} |x_k\rangle\langle v_k|\right)|x_j\rangle$ and $\| \, |y_0, \gamma\rangle \| = \sqrt{n_j - 1}/\sqrt{n_j}$. Furthermore, by eq. (2.23),

$$U|x_0, y_0\rangle = P|x_0, y_0\rangle = |y_0, x_0\rangle.$$

Therefore, in the general case where the input register is $s|x_j\rangle + t|x_0\rangle$ (with $|s|^2 + |t|^2 = 1$) we have:

$$\begin{aligned} U[(s|x_j\rangle + t|x_0\rangle) \otimes |y_0\rangle] &= sU|x_j, y_0\rangle + tU|x_0, y_0\rangle \\ &= s(s_j|f(x_j), x_0\rangle + |y_0, \gamma\rangle) + t|y_0, x_0\rangle \\ &= ss_j|f(x_j), x_0\rangle + |y_0\rangle(s|\gamma\rangle + t|x_0\rangle). \end{aligned}$$

As a consequence, the result that we want is in the first $[\mathcal{H}(\Omega^*)]$ quantum register, but its $|y_0\rangle$ component is garbage and should be ignored in subsequent computations. Furthermore, the amplitude of the desired result will decrease through successive computation stages as a result of attenuation by successive $1/\sqrt{n_j}$ factors.

As discussed previously, quantum states with $|x_j\rangle$ components with positive amplitudes represent sets of the corresponding x_j ($j \neq 0$). Applying U to such a state yields a quantum state with positive amplitudes for the corresponding $|f(x_j)\rangle$, which represents the set of corresponding outputs $f(x_j)$. Thus, U can be applied to a (crisp) set of values with quantum parallelism.

2.4 Topographic qubit maps

2.4.1 Representation

To further explore quantum computation by topographic maps, in this section we present an alternative representation of the maps and a circuit-based implementation of arbitrary functions on a finite domain. In these *topographic qubit maps*, each domain value $x_j \in \Omega = \{x_1, \ldots, x_n\}$ or codomain value $y_i \in \Omega' = \{y_1, \ldots, y_m\}$ is assigned a separate qubit, whose state, for example, $|\psi_j\rangle = p_j'|0\rangle + p_j|1\rangle$, where $|p_j'|^2 + |p_j|^2 = 1$, represents the activity level of neuron x_j by the amplitude p_j. This topographic representation can be contrasted with a binary representation such as that proposed by Chen, Wang, and Charbon [17]. Our one-out-of-n (or "one-hot") representation might seem unrealistically inefficient, but (1) we are assuming a scalable qubit implementation, which permits arrays of many thousands of qubits, and (2) neural computations typically require only low precision [in the brain perhaps as little as one digit: [18]]. Therefore a quantity can be represented by a few tens of qubits; that is, our m and n will typically be small ($m, n < 100$).

Like the topographic basis maps, these topographic qubit maps can also be viewed as representations of subsets of the domain; for $S \subseteq \Omega$:

$$|S\rangle = \sum_{x_j \in S} |1\rangle_j + \sum_{x_j \notin S} |0\rangle_j.$$

That is, the x_j qubit is in state $|1\rangle$ if x_j is in S and is in state $|0\rangle$ if it is not. Therefore we use the notation $|\{x_j\}\rangle$ for the topographic map representing just the number x_j. The set of representations of all possible subsets of Ω is then an ON basis for the 2^n-dimensional Hilbert space of these qubits. The basis can be written:

$$\{|k\rangle | k \in \mathbf{2}^n\} = \{|S\rangle | S \subseteq \mathbf{2}^\Omega\},$$

where on the left $\mathbf{2}^n$ is the set of n-bit binary strings, and on the right $\mathbf{2}^\Omega$ is the powerset of Ω. Therefore, the sets are basis states and as a consequence topographic qubit maps permit *multiple* sets to be processed in quantum superposition. Moreover, because the sets are represented by computational basis vectors, they can be copied without violating the no-cloning theorem.

By using amplitudes other than 0 and 1, we can represent fuzzy sets. Suppose S is a fuzzy set with membership function μ: $S \to [0, 1]$, and let $m_j = \mu(x_j)$. Then S is represented by the topographic qubit map

$$|S\rangle = \sum_{j=1}^{n} m_j|1\rangle_j + \sqrt{1-m_j^2}\;|0\rangle_j.$$

Fuzzy sets cannot, in general, be copied (nor can arbitrary superpositions of crisp sets).

With the topographic qubit representation, the transformations between computational maps will be implemented by the quantum circuit model, and so one might ask whether it would be simpler to implement ordinary binary digital quantum computation. The answer is that computation on topographic maps can be implemented by a few relatively simple operations (described in the following subsections), so that computational maps buy a simpler quantum implementation at the cost of greater representational expense (number of qubits). We expect topographic quantum computation to be more simply implemented than a full-scale digital quantum arithmetic-logic unit.

2.4.2 Unary functions

An example will illustrate how to implement an arbitrary finite function f: $\Omega \to \Omega'$ by a unitary operator on topographic qubit maps. For any y_i not in the range of f, we set its state $|\phi_i\rangle = |0\rangle$ supplied as an ancilla. If y_i is in the range of f, then it might be the image of a single domain element, x_j, that is, $y_i = f(x_j)$, in which case we implement directly $|\phi_i\rangle = |\psi_j\rangle$, transferring the state $|\psi_j\rangle$ of input qubit j to output qubit i. If there are two domain values mapping into y_i, say $f(x_j) = y_i = f(x_k)$, then we make $|\phi_i\rangle$ the logical OR of $|\psi_j\rangle$ and $|\psi_k\rangle$. This is accomplished with the two-input OR_2 gate:

$$\mathrm{OR}_2 \stackrel{\mathrm{def}}{=} \mathrm{CCNOT}(\mathrm{X} \otimes \mathrm{X} \otimes \mathrm{I}), \tag{2.26}$$

where CCNOT is the conditional-conditional-not or Toffoli gate. The result of ORing the input states is:

$$\mathrm{OR}_2\big(|\psi_j\rangle \otimes |\psi_k\rangle \otimes |1\rangle\big) = \mathrm{X}|\psi_j\rangle \otimes \mathrm{X}|\psi_k\rangle \otimes |\phi_i\rangle. \tag{2.27}$$

The $|1\rangle$ input is a constant ancilla. The result of the OR is in the third output qubit, and the first two output qubits, in which the negated inputs remain, are considered garbage. If $|\psi_j\rangle = p'|0\rangle + p|1\rangle$ and $|\psi_k\rangle = q'|0\rangle + q|1\rangle$, then OR_2 transfers the probability amplitudes as follows:

$$\begin{aligned}\mathrm{OR}_2|\psi_j\rangle|\psi_k\rangle|1\rangle &= \mathrm{OR}_2(p'|0\rangle + p|1\rangle)(q'|0\rangle + q|1\rangle)|1\rangle \\ &= p'q'\mathrm{OR}_2|001\rangle + p'q\mathrm{OR}_2|011\rangle + pq'\mathrm{OR}_2|101\rangle + pq\mathrm{OR}_2|111\rangle \\ &= p'q'|110\rangle + p'q|101\rangle + pq'|011\rangle + pq|001\rangle.\end{aligned}$$

Therefore, the third output qubit is the OR of the first two input qubits with the amplitudes shown. If we interpret the squares of the amplitudes as probabilities, then OR_2 computes the correct probabilities for the third output qubit. The first two output qubits are considered garbage but must be retained, for they are entangled with the third output.

If more than two domain values map into a single codomain value, then we use the multiple argument OR_n, which can be defined recursively in terms of OR_2:

$$\mathrm{OR}_n|\psi_1\rangle\cdots|\psi_n\rangle|1\rangle^{\otimes(n-1)} \stackrel{\text{def}}{=} \mathrm{OR}_{n-1}\left[(\mathrm{OR}_2|\psi_1\rangle|\psi_2\rangle|1\rangle)\otimes|\psi_3\rangle\cdots|\psi_n\rangle|1\rangle^{\otimes(n-2)}\right] \quad (n>2). \tag{2.28}$$

For completeness, we define $\mathrm{OR}_1 \stackrel{\text{def}}{=} \mathrm{I}$.

With the preceding motivation, we can give the construction for computing an arbitrary finite function by topographic qubit maps:

Proposition 5 Suppose $f:\Omega \to \Omega'$, where $\Omega = \{x_1, \ldots, x_n\}$ and $\Omega' = \{y_1, \ldots, y_n\}$. Let $m_{\mathrm{nr}} \stackrel{\text{def}}{=} n - |\mathrm{Im} f|$ be the number of codomain elements that are not in the range of f. Let n_{b} be the number of injective domain elements, and let $n_{\mathrm{n}} \stackrel{\text{def}}{=} n - n_{\mathrm{b}}$ be the number of non-injective domain elements. Let $m_{\mathrm{n}} \stackrel{\text{def}}{=} |\mathrm{Im} f| - n_{\mathrm{b}}$ be the number of non-injective range elements (i.e., those that are the image of two or more domain elements). Then there is an $2n_{\mathrm{n}} + n_{\mathrm{b}} + m_{\mathrm{nr}} - m_{\mathrm{n}}$ dimensional unitary operator U_f that computes f by topographic qubit maps as follows:

$$U_f\ |\{x\}\rangle\ |0\rangle^{\otimes m_{\mathrm{nr}}}\ |1\rangle^{\otimes(n_{\mathrm{n}} - m_{\mathrm{n}})} = |\{f(x)\}\rangle|y\rangle, \tag{2.29}$$

where $|y\rangle$ is $2(n_{\mathrm{n}} - m_{\mathrm{n}})$ qubits of garbage.

Proof: The inputs are the n elements of the input map, m_{nr} constant $|0\rangle$ ancillae (for the non-range elements), and $n_{\mathrm{n}} - m_{\mathrm{n}}$ constant $|1\rangle$ ancillae for the OR_2 gates that map multiple domain elements to the same range element. The latter are required because each of the non-injective range elements requires a number of OR_2 gates that is one less than the number of its preimages; hence the m_{n} non-injective range elements require $n_{\mathrm{n}} - m_{\mathrm{n}}$ OR_2 gates. Therefore there are

$$n + m_{\mathrm{nr}} + n_{\mathrm{n}} - m_{\mathrm{n}} = (n_{\mathrm{b}} + n_{\mathrm{n}}) + m_{\mathrm{nr}} + n_{\mathrm{n}} - m_{\mathrm{n}} = 2n_{\mathrm{n}} + n_{\mathrm{b}} + m_{\mathrm{nr}} - m_{\mathrm{n}}$$

input qubits. The m_{nr} constant $|0\rangle$s are passed directly to the output qubits for non-range codomain elements. The n_{b} qubits for injective inputs are passed to the same number of output qubits, permuted as required. The outputs of the ORs project onto the m_{n} qubits that represent range values with more than one preimage. Each OR_2 also generates two garbage qubits, for a total of $2(n_{\mathrm{n}} - m_{\mathrm{n}})$. Therefore the total number of output qubits is

$$m_{\mathrm{nr}} + n_{\mathrm{b}} + m_{\mathrm{n}} + 2(n_{\mathrm{n}} - m_{\mathrm{n}}) = 2n_{\mathrm{n}} + n_{\mathrm{b}} + m_{\mathrm{nr}} - m_{\mathrm{n}},$$

which is equal to the number of input qubits, as it should be. Next we define U_f explicitly as the tensor product of three operators:

$$U_f \stackrel{\mathrm{def}}{=} U_{\mathrm{b}} \otimes U_{\mathrm{nr}} \otimes U_{\mathrm{n}}. \tag{2.30}$$

We will use the notation $|1\rangle_q$ to represent a $|1\rangle$ state in qubit q, and $|0\rangle_q$ to represent a $|0\rangle$ state in the qubit q.

The U_{nr} operator is an identity operation copying constant $|0\rangle$ ancillae into the codomain elements that are not in the range of f. Therefore, let $\{c_1, \ldots, c_{m_{\mathrm{nr}}}\} \subseteq \Omega' - \mathrm{Im} f$ be this subset, and let $|z_i\rangle$ be ancillae qubits to provide constant $|0\rangle$s. Then $U_{\mathrm{nr}} \in \mathcal{L}(\mathcal{H}^{m_{\mathrm{nr}}}, \mathcal{H}^{m_{\mathrm{nr}}})$ is defined:

$$U_{\mathrm{nr}} \stackrel{\mathrm{def}}{=} \sum_{i=1}^{m_{\mathrm{nr}}} |0\rangle_{c_i} \langle 0|_{z_i} + |1\rangle_{c_i} \langle 1|_{z_i}. \tag{2.31}$$

That is, the states of the z_i input qubits (intended to be $|0\rangle$) are transferred to the c_i output qubits. This operator can be abbreviated by the following bracket notation:

$$U_{\mathrm{nr}} \stackrel{\mathrm{def}}{=} [c_1, \ldots, c_{m_{\mathrm{nr}}}] \leftarrow [z_1, \ldots, z_{m_{\mathrm{nr}}}]. \tag{2.32}$$

It is just a permutation of the qubits, which could be implemented by qubit SWAP operations.

The U_{b} operator handles the domain elements that are mapped injectively to their images. Therefore, let $\{v_1, \ldots, v_{n_{\mathrm{b}}}\} \subseteq \mathrm{Im} f$ be the injective domain elements, and let $u_i = f(v_i)$ be the corresponding range elements. Then $U_{\mathrm{b}} \in \mathcal{L}(\mathcal{H}^{n_{\mathrm{b}}}, \mathcal{H}^{n_{\mathrm{b}}})$ is a permutation of this subset of the topographic map elements:

$$U_{\mathrm{b}} \stackrel{\mathrm{def}}{=} \left[u_1, \ldots, u_{n_{\mathrm{b}}}\right] \leftarrow \left[v_1, \ldots, v_{n_{\mathrm{b}}}\right]. \tag{2.33}$$

For U_{n} we must OR together the domain elements that correspond to each range element with more than one preimage. Therefore we define U_{n} as a tensor product of operators for each such range element:

$$U_{\mathrm{n}} \stackrel{\mathrm{def}}{=} \bigotimes_{i=1}^{m_{\mathrm{n}}} V_i\left(y_i, f^{-1}\{y_i\}\right), \tag{2.34}$$

where these y_i each have more than one preimage element; for example, $f^{-1}\{y_i\} = \{x_1, \ldots, x_{n_i}\}$, where $n_i = |f^{-1}\{y_i\}| \geq 2$. The output state $|\psi_i\rangle$ for such a range element is the OR of the input states $|\xi_j\rangle$ $(j = 1, \ldots, n_i)$ of its preimage elements:

$$|\psi_i\rangle|\gamma\rangle = \mathrm{OR}_{n_i}|\xi_1\rangle \cdots |\xi_{n_i}\rangle|1\rangle^{\otimes(n_i-1)},$$

where OR_{n_i} is a cascade of $n_i - 1$ OR_2 gates and $|\gamma\rangle$ is $2(n_i - 1)$ dimensional garbage. This is accomplished by the operators $V_i(y_i, \{x_1, \ldots, x_{n_i}\}) \in \mathcal{L}(\mathcal{H}^{2n_i-1}, \mathcal{H}^{2n_i-1})$:

$$V_i(y_i, \{x_1, \ldots, x_{n_i}\}) \stackrel{\text{def}}{=} \left[y_i, \gamma_1, \ldots, \gamma_{2(n_i-1)}\right] \leftarrow \mathrm{OR}_{n_i}\left[x_1, \ldots, x_{n_i}; o_1, \ldots, o_{n_i-1}\right], \tag{2.35}$$

where $o_1, \ldots, o_{n_i-1}$ are the qubits that provide ancillary $|1\rangle$s for the ORs, and the garbage outputs $\gamma_1, \ldots, \gamma_{2(n_i-1)}$ receive the negated inputs and intermediate OR outputs. The bracket notation identifies the qubits that are the inputs and outputs of OR_{n_i}. This completes the construction of U_f [eq. (2.29)].

There are more efficient ways to compute U_f, but the above construction is easier to understand. This basic approach can be used to approximate a variety of unary functions useful in neural networks, such as sigmoid functions, including non-injective, non-surjective squashing functions. However, neural networks also require non-unary functions such as addition and multiplication, to which we now turn.

2.4.3 Binary functions

In sensory cortical areas there are many topographic maps that represent two or more dimensions of a stimulus (e.g., retinal position and edge orientation); localized activity in these maps represent conjunctions of values on these dimensions. Similarly, quantum computational maps can represent conjunctions of values as inputs or outputs of functions.

Suppose we want to compute a function $f{:}\Omega \times \Omega \rightarrow \Omega'$, where $\Omega = \{x_1, \ldots, x_n\}$ and $\Omega' = \{y_1, \ldots, y_n\}$. We will represent the input to the function by a two-dimensional array of qubits for each (x_j, x_k) pair. (They do not have to be physically arranged as a two-dimensional array so long as there is a qubit for each pair of values.) This will require n^2 qubits, but we are assuming that n is small because low precision is adequate for neural networks. Therefore we expect the 2D map to comprise typically several thousand qubits. The qubits representing the (x_j, x_k) pairs are then mapped to the qubits representing the outputs $f(x_j, x_k)$ by the method described in Prop. 5.

The n^2 conjunctions are computed by n^2 CCNOT gates, each of which requires a $|0\rangle$ ancilla and generates two extra qubits (containing the inputs) in addition to the conjunction. However, these extra qubits are passed along the rows and columns to be used in other conjunctions, and so there are only $2n$ total garbage qubits. In summary, there are $2n+n^2$ input qubits (including n^2 ancillae) and n^2+2n output qubits (including $2n$ garbage). That is, if $|\phi_j\rangle$ is the state of element j of one input map, and $|\psi_k\rangle$ is the state of element k of the other input map, then the state $|\chi_{jk}\rangle$ of element (j, k) of the two-dimensional map is computed by

$$|\phi_j\rangle|\psi_k\rangle|\chi_{jk}\rangle = \text{CCNOT}|\phi_j\rangle|\psi_k\rangle|0\rangle. \tag{2.36}$$

If $|\phi_j\rangle = p'|0\rangle + p|1\rangle$ and $|\psi_k\rangle = q'|0\rangle + q|1\rangle$, then the result of CCNOT is

$$|\phi_j\rangle|\psi_k\rangle|\chi_{jk}\rangle = p'q'|000\rangle + p'q|100\rangle + pq'|010\rangle + pq|111\rangle.$$

The qubits are entangled, but the conjunction computes probability-like amplitudes if we interpret the squares of the amplitudes as probabilities.

Based on the foregoing, we define a unitary operator U_{OP} on a $n^2 + 2n$ dimensional Hilbert space that does what amounts to an outer product on two one-dimensional maps to compute a two-dimensional map:

$$|\varphi\rangle|\psi\rangle|\chi\rangle = U_{\text{OP}}\ |\varphi\rangle|\psi\rangle|0\rangle^{\otimes n^2}, \tag{2.37}$$

where inputs $|\varphi\rangle$ and $|\psi\rangle$ are n-dimensional, and output $|\chi\rangle$ is n^2-dimensional and computed by eq. (2.36).

To illustrate the use of computational maps to implement binary operations, we will use a simple, useful function, addition. We want to define $\text{sum}:\Omega\times\Omega \to \Omega'$ so that $\text{sum}(x,\ y) = x+y$, but we have a problem, since the maximum value of $x+y$ is greater than the maximums of x and y. Since the range of numbers represented by our maps is quite limited, this is a more serious problem than overflow in binary addition. One solution is to make the codomain map large enough; for example, if $\Omega = \{0, \Delta x, 2\Delta x,\ \ldots,\ (n-1)\Delta x\}$, then let $\Omega' = \{0, \Delta x, 2\Delta x,\ \ldots, 2(n-1)\Delta x\}$. Generally, however, it is more convenient to have the codomain map be the same as the domain maps, since this facilitates composing functions. Therefore, another solution is to scale either the inputs or the output so that we compute, for example, $\text{hsum}(x,\ y) = (x+y)/2$; this is often useful if we know that we are going to scale the quantities anyway. A third option is to compose addition with a squashing function, so that we compute, for example, the truncated $\text{tsum}(x,\ y) = \min(x+y,\ x_n)$, where $x_n = \max\Omega$. This is the solution that we will use for illustration.

If $\Omega = \{0, \Delta x,\ \ldots, (n-1)\Delta x\}$, then the (j, k) element of the two-dimensional qubit map will represent the pair of inputs $((j-1)\Delta x, (k-1)\Delta x)$. This will be mapped to the sum $(j+k-2)\Delta x$ if $j+k-2<n-1$, and to the maximum value $(n-1)\Delta x$ otherwise. Therefore the constant $j+k$ anti-diagonals above the $j+k=n+1$ anti-diagonal each map to one value, $(j+k-2)\Delta x$, which is the sum of $(j-1)\Delta x$ and $(k-1)\Delta x$. The elements below the $j+k=n+1$ anti-diagonal all map to the same $(n-1)\Delta x$

value, since the sum is above the maximum representable value in the topographic map.

Proposition 5 shows how to implement the truncated addition, but it treats it as a unary function on an n^2-dimensional space, which is wasteful since the intended output (the sum) is n-dimensional and the remaining $n^2 - n$ elements are garbage. Therefore, we implement a unitary operator that directly maps the input pairs to the corresponding outputs.

To compute the outer product we require n^2 constant $|0\rangle$ ancillae, and this computation also passes the two n-dimensional inputs through as garbage output. The qubit representing (0, 0) maps bijectively to the output qubit y_1 representing 0. Each of the other $n-1$ output qubits y_i $(i=2, \ldots, n)$ has two or more domain pairs mapping to it. As before, let n_i be the preimage multiplicity of output i and note that $\sum_{i=1}^{n} n_i = n^2$. Each of these non-injective outputs receives its value from an OR_{n_i} operation, which requires $n_i - 1$ input $|1\rangle$ ancillae and generates $2n_i - 2$ qubits of output garbage (n_i for the negated inputs and $n_i - 1$ for the intermediate disjunctions). Therefore, the total number of $|1\rangle$ ancillae is

$$\sum_{i=2}^{n}(n_i - 1) = \sum_{i=2}^{n} n_i - (n-1) = \left(n^2 - 1\right) - n + 1 = n^2 - n.$$

Moreover, the complete input dimension is $2n + n^2 + n^2 - n = 2n^2 + n$.

This is also the complete output dimension, for we have n qubits for the function value, $2n$ qubits for the passed-through input arguments (garbage), and the garbage output from the OR gates, which is:

$$\sum_{i=2}^{n}(2n_i - 2) = 2\left(n^2 - n\right).$$

That is, the complete output dimension is $3n + 2(n^2 - n) = 2n^2 + n$.

In summary, there is a unitary operator $U_{\mathrm{tsum}} \in \mathcal{L}\left(\mathcal{H}^{2n^2+n}, \mathcal{H}^{2n^2+n}\right)$ so that

$$U_{\mathrm{tsum}}\ |\{x\}\rangle\ |\{y\}\rangle\ |0\rangle^{\otimes n^2}\ |1\rangle^{\otimes n(n-1)} = |\{\mathrm{tsum}(x,\ y)\}\rangle\ |\{x\}\rangle\ |\{y\}\rangle\ |\gamma\rangle, \tag{2.38}$$

where the garbage $|\gamma\rangle$ has dimension $2n(n-1)$ (the passed-through inputs may also be considered garbage).

Based on this example, we state a more general result.

Proposition 6 Suppose $f{:}\Omega \times \Omega \to \Omega$, and let $n = |\Omega|$. Let $m_{\mathrm{nr}} \stackrel{\mathrm{def}}{=} n - |\mathrm{Im}\, f|$ be the number of codomain elements that are not in the range of f. Then there is a unitary operator $U_f \in \mathcal{L}(\mathcal{H}, \mathcal{H})$, where $\mathcal{H}$ is $2n^2 + n + 2m_{\mathrm{nr}}$ dimensional Hilbert space, such that:

$$U_f\ |\{x\}\rangle\ |\{y\}\rangle\ |0\rangle^{\otimes\left(n^2 + m_{\mathrm{nr}}\right)}\ |1\rangle^{\otimes\left(n^2 - n + m_{\mathrm{nr}}\right)} = |\{f(x,\ y)\}\rangle\ |\{x\}\rangle\ |\{y\}\rangle\ |\gamma\rangle, \tag{2.39}$$

where the garbage $|\gamma\rangle$ has dimension $2(n^2 - n + m_{\mathrm{nr}})$ (the $2n$ passed-through inputs may also be considered garbage).

Proof: The operator is constructed very similarly to U_{tsum}, but we also have to consider non-range codomain elements for non-surjective functions, which didn't occur in that case. As before, the computation of the outer product will require n^2 ancillary $|0\rangle$ inputs and it will generate $2n$ qubits containing the passed-through inputs. We can consider disjoint subsets of the codomain. Codomain elements that are not in the range of f will need to be sent a $|0\rangle$ state, for which we need an additional m_{nr} ancillary $|0\rangle$ inputs. Let n_{b} be the number of input pairs that are mapped one-to-one to the corresponding outputs; they neither require ancillary constants nor generate garbage. The remaining codomain elements are range elements with $n_i \geq 2$; let $m_{\text{n}} = n - m_{\text{nr}} - n_{\text{b}}$ be the number of them. Each of these will receive the OR of the corresponding (preimage) domain elements. As we saw previously, the OR_{n_i} operation requires $n_i - 1$ ancillary $|1\rangle$ qubits and produces $2n_i - 2$ qubits of garbage. Therefore, the total number of $|1\rangle$ qubits required for the $n^2 - n_{\text{b}}$ input pairs mapping to the m_{n} non-injectively mapped range elements is:

$$\sum_{i=1}^{m_{\text{n}}} (n_i - 1) = \sum_{i=1}^{m_{\text{n}}} n_i - m_{\text{n}} = n^2 - n_{\text{b}} - m_{\text{n}} = n^2 - n + m_{\text{nr}}, \tag{2.40}$$

since $m_{\text{n}} = n - m_{\text{nr}} - n_{\text{b}}$. The garbage generated by the ORs is then

$$\sum_{i=1}^{m_{\text{n}}} (2n_i - 2) = 2\left(n^2 - n + m_{\text{nr}}\right). \tag{2.41}$$

The complete input dimension is $2n$ (arguments) + $(n^2 + m_{\text{nr}})$ (for $|0\rangle$ ancillae) + $(n^2 - n + m_{\text{nr}})$ [for $|1\rangle$ ancillae, eq. (2.40)] $= 2n^2 + n + 2m_{\text{nr}}$. The output dimension is $3n$ (arguments and result) + $2(n^2 - n + m_{\text{nr}})$ [garbage, eq. (2.41)] $= 2n^2 + n + 2m_{\text{nr}}$.

The same approach can be used for operations with more than two arguments, but the number of qubits increases exponentially with the number of arguments.

2.5 Conversions between representations

Ordinary binary representations can be translated to topographic qubit maps by a unitary demultiplexer U_{demux} that operates on an m-qubit binary number $|k\rangle$ and directs a $|1\rangle$ qubit to the kth of $n = 2^m$ output qubits (the remainder receiving $|0\rangle$). Let $|\{k\}\rangle$ be the resulting computational qubit map. Then:

$$U_{\text{demux}}|k\rangle|1\rangle|0\rangle^{\otimes(n-1)} = |k\rangle|\{k\}\rangle. \tag{2.42}$$

U_{demux} operates on an $m + n = m + 2^m$ dimensional Hilbert space. A demultiplexer can be implemented with CSWAP (Fredkin) gates [19].

The opposite translation, from a computational map to a binary representation, is more complicated. First, we must decide what we want it to do, for in general a topographic qubit map represents multiple values with different amplitudes, $|\psi\rangle = \bigotimes_{j=1}^{n}(p_j'|0\rangle + p_j|1\rangle)$, where $|p_j'|^2 + |p_j|^2 = 1$. Which x_j should it produce? The one with the maximum amplitude? (And what if several have the same maximum amplitude?) An x_j chosen probabilistically based on $|p_j|^2$? The binary representation of a weighted average, $n^{-1}\sum_{j=1}^{n} p_j x_j$? A normalized superposition of all the values? The answer is not apparent, so we leave the question open.[4]

2.6 Applications to quantum machine learning

Given this general ability to compute non-unitary and even nonlinear functions by means of topographic qubit maps, it is possible to do the operations useful for machine learning such as inner products and sigmoid nonlinearities. For example, an inner product of N-dimensional vectors requires N multiplications and $N-1$ additions. If $|\Omega| = n$, then each multiplication and addition will require approximately $2n^2$ qubits (Prop. 6), for a total of approximately $2N^2n^2$.

For one layer of a neural network, with say N neurons projecting through an $M \times N$ weight matrix into M neurons, we must do M inner products with the input. Since crisp sets are represented by computational maps that are basis vectors in the computational basis, they can be copied. Therefore, the N-dimensional input vector can be copied $M-1$ times to do the M inner products. (This requires $M-1$ CNOT gates and $(M-1)n$ ancillary $|0\rangle$ qubits. Overall, one layer requires about $2MN^2n^2$ qubits for the computation (not including ancillary qubits).

2.7 Conclusions

Topographic (computational) maps are widely used in the brain to implement simultaneous nonlinear vector transformations on multiple inputs. In this chapter we have explored two approaches to quantum topographic computing with a focus on brain-inspired machine learning applications. The first, called a topographic basis map, assigns locations in the map to state vectors in a continuous or discrete basis for a quantum Hilbert space. Arbitrary functions can be implemented on such maps, which can be interpreted as representing crisp sets of inputs, but there is an unavoidable data-dependent attenuation of the result (relative to a "garbage" state) that is not easily avoidable. The second approach, called a topographic qubit map, assigns a separate qubit to each location in the map and uses the relative amplitude

4 It is easy however to produce the binary representation of either the maximum or minimum number with unit amplitude ($p_j = 1$, $p_j' = 0$) in a map.

of the $|1\rangle$ and $|0\rangle$ states to represent the presence of values in the (crisp or fuzzy) set represented by the map. Arbitrary functions on these maps are implemented by well-known quantum logic gates. In particular, computational maps enable the implementation of the functions commonly used in artificial neural networks.

References

[1] Nielsen, M. A. and Chuang, I. L. 2010. *Quantum computation and quantum information* (10th Anniversary Edition ed.). Cambridge, UK: Cambridge Univ. Press.

[2] Schuld, M., Sinayskiy, I., and Petruccione, F. 2014. The quest for a quantum neural network. *Quantum Information Processing* 13:2567–2586. https://doi.org/10.1007/s11128-014-0809-8

[3] Schuld, M., Sinayskiy, I., and Petruccione, F. 2015. An Introduction to quantum machine learning. *Contemporary Physics* 56:172–185. https://doi.org/10.1080/00107514.2014.964942

[4] Furber, E. F., Lester, D. R., Plana, L. A., Garside, J. D., Painkras, E., Temple, S., and Brown, A. D. 2013. Overview of the SpiNNaker system architecture. *IEEE Transactions on Computers* 62:2454–2467. https://doi.org/10.1109/TC.2012.142

[5] Neftci, E., Binas, J., Rutishauser, U., Chicca, E., Indiveri, G., and Douglas, R. J. 2013. Synthesizing cognition in neuromorphic electronic systems. *Proceedings of the National Academy of Sciences of the United States of America* 110:E3468–E3476. http://dx.doi.org/10.1073/pnas.1212083110

[6] O'Reilly, R. C., Munakata, Y., Frank, M. J., Hazy, T. E., and contributors. 2014. *Computational cognitive neuroscience* (2nd ed.). http://ccnbook.colorado.edu/ (accessed January 12, 2019)

[7] MacLennan, B. J. 2017. Field computation: A framework for quantum-inspired computing. In *Quantum Inspired Computational Intelligence: Research and Applications*, ed. S. Bhattacharyya, U. Maulik, and P. Dutta, 85–110. Cambridge, MA: Morgan Kaufmann / Elsevier.

[8] Morasso, P. G., and Sanguineti, V. 1997. *Self-organization, computational maps and motor control*. Amsterdam: North-Holland.

[9] MacLennan, B. J. 1999. Field computation in natural and artificial intelligence. *Information Sciences* 119:73–89.

[10] Eliasmith, C., Stewart, T. C., Choo, X., Bekolay, T., DeWolf, T., Tang, Y., and Rasmussen, D. 2012. A large-scale model of the functioning brain. *Science* 338:1202–1205. http://doi.org/10.1126/science.1225266

[11] MacLennan, B. J. 1997. Field computation in motor control. In *Self-Organization, Computational Maps and Motor Control*, ed. P. G. Morasso and V. Sanguineti, 37–73. Amsterdam: North-Holland.

[12] MacLennan, B. J. 2009. Field computation in natural and artificial intelligence. In *Encyclopedia of Complexity and System Science*, ed. R. Meyers et al., Chapter 6, entry 199, 3334–3360. New York: Springer.

[13] MacLennan, B. J. 2014. The promise of analog computation. *International Journal of General Systems* 43:682–696.

[14] Rumelhart, D. E., McClelland, J. L., and PDP Research Group. 1986. *Parallel distributed processing: Explorations in the microstructure of cognition, volume 1: Foundations*. Cambridge, MA: MIT Press.

[15] Sanger, T. D. 1996. Probability density estimation for the interpretation of neural population codes. *Journal of Neurophysiology* 76:2790–2793.

[16] Høyer, P. 2000. Arbitrary phases in quantum amplitude amplification. *Physical Review A* 62:052304. https://doi.org/10.1103/PhysRevA.62.052304
[17] Chen, J., Wang, L., and Charbon, E. 2017. A quantum-implementable neural network model. *Quantum Information Processing* 16:245. http://doi.org/10.1007/s11128-017-1692-x
[18] McClelland, J. L., Rumelhart, D. E., and PDP Research Group. 1986. *Parallel distributed processing: Explorations in the microstructure of cognition, volume 2: Psychological and biological models*. Cambridge, MA: MIT Press.
[19] Fredkin, E. F. and Toffoli, T. 1982. Conservative logic. *International Journal of Theoretical Physics* 21:219–253.

Ashish Mani, Siddhartha Bhattacharyya, and Amlan Chatterjee

3 Quantum optimization for machine learning

Abstract: Machine learning is a branch of Artificial Intelligence that seeks to make machines learn from data. It is being applied for solving real world problems with huge amount of data. Though, Machine Learning is receiving wide acceptance, however, execution time is one of the major concerns in practical implementations of Machine Learning techniques. It largely comprises of a set of techniques that trains a model by reducing the error between the desired or actual outcome and an estimated or predicted outcome, which is often called as loss function. Thus, training in machine learning techniques often requires solving a difficult optimization problem, which is the most expensive step in the entire model-building process and its applications. One of the possible solutions in near future for reducing execution time of training process in Machine learning techniques is to implement them on quantum computers instead of classical computers. It is conjectured that quantum computers may be exponentially faster than classical computers for solving problems which involve matrix operations. Some of the machine learning techniques like support vector machines make extensive use of matrices, which can be made faster by implementing them on quantum computers. However, their efficient implementation is non-trivial and requires existence of quantum memories. Thus, another possible solution in near term is to use a hybrid of Classical Quantum approach, where a machine learning model is implemented in classical computer but the optimization of loss function during training is performed on quantum computer instead of classical computer. Several Quantum optimization algorithms have been proposed in recent years, which can be classified as gradient based and gradient free optimization techniques. Gradient based techniques require the nature of optimization problem being solved to be convex, continuous and differentiable otherwise if the problem is non-convex then they can find local optima only whereas gradient free optimization techniques work well even with non-continuous, non-linear and non-convex optimization problems. This chapter discusses a global optimization technique based on Adiabatic Quantum Computation (AQC) to solve minimization of loss function without any restriction on its structure and the underlying model, which is being learned. Further, it is also shown that in the proposed framework, AQC based approach would be superior to circuit-based approach in solving global optimization problems.

Keywords: Artificial Intelligence, quantum computing, non-convex optimization

Ashish Mani, ASET, Amity University UP, Noida, Uttar Pradesh, India
Siddhartha Bhattacharyya, Department of Computer Science and Engineering, CHRIST (Deemed to be University), Bangalore, India
Amlan Chatterjee, Computer Science, California State University, Dominguez Hills, USA

https://doi.org/10.1515/9783110670707-003

3.1 Introduction

Humanity has been enamored with the creation of artificially intelligent machines since the advent of programmable electronic computing devices in last century. In 1956, the term 'Artificial Intelligence' was coined for 'thinking machines' in the workshop held at Dartmouth [1]. Several proposals have been made to make machines behave intelligently [2], but most of them suffered from non-availability of technology required to implement them efficiently. Learning is considered as manifestation of the intelligent behavior, so a machine which can learn a difficult task will be treated at least artificially intelligent. The discipline which deals with the learning techniques involving data is called as Machine learning. It tries to make a computer learn from past experiences represented in from of data and make reliable decisions from it. These techniques can fine tune their performance as fresh experiences are gathered. It helps in building a data driven model rather than a static precise model and is widely used in classification, clustering and pattern recognition etc. It is being widely used in real world fields like financial services [3], health services [4], marketing and sales [5], social network analysis [6], and virtual personal assistants [7] etc.

Some popular Machine Learning algorithms are Artificial Neural Networks (ANN) [8], Support Vector Machines (SVM) [9], K-means Clustering [10], Regression [11], Decision trees [12] etc. The generic steps involved in any machine learning project are as follows:

Data collection: This step involves collecting data pertaining to the problem to be solved. The data should be of high quality, which is necessary for a high quality model. There are several measures available to evaluate the quality of data [13]. Low quality of data will generally result in low quality of model and subsequently predictions or decisions.

Model selection: This step involves selection of a specific technique like ANN or SVM etc. Further, it also involves selecting the structure of the model i.e. in case of ANN, number of hidden layers, no. of neurons in each hidden layer, activation functions etc.

Training: In this step, the parameters of the model are decided i.e. in case of ANN and SVM, weights are determined using some loss function i.e., a loss function which represents the performance of the model on training data set is used to search / compute the weights.

Validation: In this step, the performance of the model on testing data set is determined and compared with other existing models.

Deployment: Finally the model is deployed so that it can be used for making predictions or decisions in real situations.

The first step is specific to the problem being solved by using machine learning, but can be automated in most cases, to make data collection process reliable, quick and cheap. The second step of Model selection usually requires past experience and trial and error approach, which can be translated into a heuristic [13]. The selection of the model is often done more than once and is typically based on the empirical evidence regarding the performance of the model, which is known after the validation. The third step of training is usually the most compute intensive approach of all the other steps and depending on the complexity of model, it can be efficient also if the underlying process is linear, then its modelling results in well-behaved loss function i.e., loss function is convex, continuous and differentiable in the region of interest. However, most loss functions are often not well behaved due to the underlying nonlinear process which is generating the data and in such cases, the optimization of loss function becomes a key challenge in successful implementation of machine learning technique. It has been often experienced that when loss function is well-behaved, the model is less useful [14] and when loss function is difficult to optimize, then the resulting model is more powerful [15]. Validation and deployment are usually computationally less expensive as the trained model takes input and gives output in the form of predictions or decisions. In this chapter, we will primarily focus on making training more efficient in machine learning through quantum optimization.

Optimization is applied routinely in almost all types of machine learning techniques and its applications. Chambers dictionary describes optimization as an act of 'making the most or best of anything'. Therefore, performing optimization in training stage of machine learning means finding the most or best suitable parameters of the model. There are machine learning models like linear SVM whose weights can be determined by matrix inversion techniques [16], which can be implemented efficiently in case of gate model of quantum computation, thus showing exponential speed-up. However, such techniques often require existence of quantum RAM [17] and data to be generated from a linear process with low noise. In real world problems, the assumption of linearity is often violated and thus most popular techniques are non-linear and convexity cannot be guaranteed. If the optimization problem is convex then gradient based techniques are guaranteed to find the global optima regardless of the initialization [18]. So many optimization techniques try to simplify the problem so that they appear linear or convex [13]. Alternatively, they try to solve the problem approximately by using techniques like stochastic gradient descent [19], which are iterative and not the desired single shot techniques. Thus, these techniques are usually computationally time-consuming as the nature of the problem they are trying to solve is non-linear, non-differentiable, multi-dimensional, multi-modal, and stochastic. Several quantum enhanced approaches have been proposed to solve computational issues in machine learning, of which global optimization using adiabatic quantum computation (AQC) is a promising technique, especially for problems which are non-linear, non-differentiable, multi-dimensional, and multi-modal. Efforts have also been made to solve quadratic unconstrained binary optimization problems on AQC model [20]. In this chapter, we propose an adiabatic quantum computing based iterative global optimizer for minimizing the loss function of machine learning techniques, which would be useful when the optimization problem is non-linear, non-differentiable, multi-dimensional, and multi-modal, where gradient based techniques like [19] and single shot techniques [21] are not computationally efficient or relevant. Further, there are some Variational Quantum computing proposals, where they are focusing on implementing function evaluation on a near term quantum computing device and implementing a traditional gradient free optimization technique like Nelder-Meads method on classical computer [19], however, our proposal is different from such proposals as the entire optimization step will be run on a future Adiabatic Quantum Computer. This chapter is further organized as follows: Related work about the development of AQC based algorithms is discussed in Section 3.2. The theoretical underpinning of Adiabatic Quantum Computing is discussed in Section 3.3. Section 3.4 presents the development of Grover's Adaptive Search. Problem formulation is described in Section 3.5. Section 3.6 presents quantum search based on AQC. Minimum finding algorithm based on AQC is described in Section 3.7. Section 3.8 discusses the framework for implementing quantum optimization algorithm based on AQC. Conclusions and some directions for future work are presented in Section 3.9.

3.2 Related work

Adiabatic quantum algorithms were introduced by Farhi et al. [22] based on the adiabatic conditions derived from Folk theorem. These algorithms have a Hamiltonian constructed for the problem being solved which is called as Problem Hamiltonian (H_{fin}). The ground state of the H_{fin} can be expressed as the optimal solution of the problem. Further, initially the system is prepared in the ground state of an initial Hamiltonian, which is easy to build. Subsequently, the initial Hamiltonian is changed progressively towards the H_{fin} so that the probability of finding the solution is large at the end of the evolution. However, the change from initial Hamiltonian to final Hamiltonian i.e. H_{fin} cannot be instantaneous or even lesser than the minimum time required according to the adiabatic theorem. If the system is evolved slowly enough then adiabatic theorem guarantees that system will remain in the ground state of the instantaneous Hamiltonian, thus the evolution time of varying the Hamiltonian from initial to problem is a measure of complexity of the AQC based algorithm. Several constraint satisfaction problems along with optimization problems can be represented by an aggregated set of local Hamiltonians $H = \sum i \ H \ i$ where Hi represents a local constraint. The ground state of H would satisfy the maximum possible number of such constraints and represents a desired optimal solution amongst other possible optimal solutions in case of multimodal problems, i.e. the problems having more than one optima solutions. The system stays in the instantaneous ground state of the evolving Hamiltonian as it will remain close to the ground state provided the rate of change of Hamiltonian from initial to H_{fin} remains within the bounds specified by the adiabatic theorem. The bounds on the evolution time is dependent on the energies of the initial and problem Hamiltonian as well as on the inverse gap of the instantaneous Hamiltonians H(t). The energy difference between the ground state of a Hamiltonian and its first excited state represents the gap of the Hamiltonian. That is when Hamiltonian is presented as a matrix then the gap is the difference between Hamiltonian's smallest and second smallest eigenvalue. Farhi et al. [22] have proposed and investigated AQC based algorithms for NP-complete problems like 3SAT. However, the key challenge in AQC based algorithm for solving a specific problem is analytical calculation of the Hamiltonian gap, which is not known for most of the problems till date. The other approach is to numerically simulate the system but the computational cost of simulating quantum mechanical system on classical computers is very expensive so the number of qubits in numerical simulations is limited. However, the numerical simulation performed in [22] on small sized problems did show that the time required to solve NP-complete problems scaled only polynomially with the problem size. But later work [23] gave strong negative evidence for the speed up shown in [22] for AQC based algorithms. It was shown that the performance of adiabatic algorithms is even sensitive to the choice of initial Hamiltonian and it can fail if the selection of initial Hamiltonian is not performed according to the structure of the problem [23]. Further, some investigations were made in [24] by mapping the H_{fin} of certain instances of 3-SAT to an Ising model and it was shown that the Hamiltonian gap is exponentially small in some problem instances. Further, some special instances of 3-SAT were also constructed in [25] which were hard for the AQC based algorithm. Subsequently, it was shown in [26, 27] that local minimum of some optimization problems can cause very small gaps in the spectrum of the Hamiltonian. These studies had designed hard instances of problems to highlight the weakness of AQC based algorithms. But in [28], randomly generated instances of the NP-complete problem Exact Cover 3 (EC3), also known as 1-in-3 SAT were used to test AQC based algorithm and it was shown that failure probability of AQC based algorithm was high. Altshuler et al. argued that their study provided strong evidence against adiabatic quantum optimization, i.e. hard instances of NP-complete problems are difficult for AQC based algorithm but the basis for this conclusion depends on some common properties shared by other NP-complete problems such as 3-SAT. However, initially Choi [29] and later Dickson and Amin [30] challenged the argument against AQC in [28]. Choi's argued that a failure of a specific instance of AQC based algorithm is not sufficient to conclude that a problem is hard for adiabatic quantum computation / optimization in general. An AQC based algorithm comprise of three variable components, namely, initial Hamiltonian, problem Hamiltonian and evolution path. Thus, there are many combinations possible of these three components for designing an AQC based algorithm. He further argued that if it is to be proved that adiabatic quantum computation is inefficient even for a particular problem then, it has to be shown that there exists no polynomial-time adiabatic quantum algorithms for the problem, which requires testing all combinations possible for designing AQC based algorithms and not just few instances. Further, Dickson and Amin [30] analytically showed that there always exist adiabatic paths for the NP-hard problem of maximum independent set along which no such crossings of local minima of the final Hamiltonian with its global minimum occur, thus the Hamiltonian gap need not be vanishingly small and hence designing of efficient AQC based algorithm is possible. Therefore, in order to prove that adiabatic quantum optimization fails for any NP-complete problem, one must prove that it is impossible to find any such path along with combination of initial and problem / final Hamiltonian in polynomial time. Therefore, it is still an open problem whether AQC can solve NP complete problems efficiently. However, it is becoming increasingly clear that there may not exit one unique path for solving all the NP complete problems.

Quantum unstructured search has been implemented adiabatically by Farhi et al. [31]. This implementation was based on global evolution model and had a computational complexity of O(N), where N is the size of the database. Later Roland and Cerf showed that quadratic speed up is possible in adiabatic version also by continuously adjusting the rate with which the initial Hamiltonian is switched to final Hamiltonian by only fulfilling locally, the conditions of adiabaticity [32]. The adiabatic quantum search has been rigorously studied by Rezakhani et al. [33] where the results relating the accuracy and the run time of the AQC based algorithm has been determined. It has also been demonstrated that AQC based algorithm shows two discernible regimes with respect to error i.e., the error reduces exponentially for short times and then reduces polynomially for longer times, under fairly general conditions.

In [34], it was demonstrated that a partial adiabatic search algorithm on quantum circuit model could be implemented and the performance of the algorithm was same as that of the local adiabatic algorithm for the case where a single element is marked i.e. $M = 1$ by investigating the minimum energy gap between the first excited state and the ground state of the system Hamiltonian. Moreover, the algorithm keeps the advantages of global adiabatic algorithms without losing the speedup of the local adiabatic search algorithm by evolving globally only for a small interval during evolution from initial to final / problem Hamiltonian.

In [22], it was shown that quantum circuit models can simulate efficiently an arbitrary AQC based algorithm. Subsequently, it has also been shown in [35] that an appropriate AQC based algorithm can efficiently simulate any quantum circuit model. Further a simple proof of equivalence between adiabatic quantum computation and quantum computation in the circuit model has been provided by Mizel et al. [36]. Therefore, AQC and Quantum Circuit models of computation are essentially equivalent, which implies that a quantum algorithm can be designed in either of the two models for solving difficult problems. Brady and Van Dam has also investigated the necessary run-time required for AQC based optimization [37] and showed that nonadiabatic speedup does not occur for general cases but only for special case.

Recently, Quantum Approximate Optimization Algorithm (QAOA) [38, 39] and the Variational Quantum Eigensolver [40] have been proposed to address classical combinatorial optimization and quantum chemistry problems, respectively.

In this chapter, a framework based on Adiabatic Quantum Computation for solving global optimization problem has been proposed in Section 3.4., which would be helpful in training of Machine Learning techniques which models nonlinear processes unlike their computationally efficient counterparts. Further, constrained optimization using AQC was discussed in [41].

3.3 Adiabatic quantum computation

Adiabatic quantum computation [42] is an alternative approach to quantum computing that has been shown to be equivalent to the quantum circuit model, but appears to have some advantages over the circuit model [31]. It is based on a well-known technique in quantum mechanics, called adiabatic approximation [43]. It is used for determining approximate solutions of Schrodinger equation subject to the condition that the Hamiltonian governing the dynamics of the system is changing slowly with time. This implies that if a quantum system is prepared in its nth eigenstate, then it will stay in the nth eigenstate throughout the evolution provided the neighborhood energy gaps are large enough and its Hamiltonian varies slowly enough.

Let a quantum system in state $|\varphi\rangle$ evolve according to the Schrodinger's equation:

$$i\hbar\frac{d}{dt}|\Phi(t)\rangle = H(t)|\Phi(t)\rangle \tag{3.1}$$

where H(t) is the Hamiltonian of the quantum system and ħ is reduced Planck's constant [31]. Let H_{ini} be the Hamiltonian of the system in its initial state i.e. at t = 0 and H_{fin} be the Hamiltonian of the system in its final state i.e. at t = T, where

$$H(t) = (1 - s(t))\ H_{ini} + s(t)H_{fin} \tag{3.2}$$

where s(t) is a continuous function with s(0) = 0 and s(T) = 1. The instantaneous eigenstates associated with H_i are denoted by $|\psi_i\rangle$. In general, let eigenstates of H(t) be $|\psi_k(t)\rangle$ such that:

$$H(t)|\psi_k(t)\rangle = E_k(t)|\psi_k(t)\rangle \tag{3.3}$$

where $E_k(t)$ are the corresponding eigenvalues and k labels the eigenstates (k = 0 indicates the ground state).

The minimum gap between the lowest two eigenvalues is defined as follows:

$$g_{min} = \min_{0\le t\le T}[E_1(t) - E_0(t)] \tag{3.4}$$

The maximum value of the matrix element of dH/dt between the two corresponding eigenstates is given by the following equation:

$$D_{max} = \max_{0 \le t \le T} \left| \left\langle \frac{dH}{dt} \right\rangle_{1,0} \right| \tag{3.5}$$

where $\left\langle \frac{dH}{dt} \right\rangle_{1,0} = \langle \psi_1(t) | \frac{dH}{dt} | \psi_0(t) \rangle$

The adiabatic theorem states that if the system at time $t = 0$ is in its ground state $|\psi_{\text{ini}}\rangle = |\psi_0(0)\rangle$ and let it evolve under the Hamiltonian $H(t)$, then

$$\left| \langle \psi_0(T) | \Phi(T) \rangle \right| \ge 1 - \varepsilon^2 \tag{3.6}$$

provided:

$$\frac{D_{max}}{g_{min}^2} \le \varepsilon \tag{3.7}$$

where $0 < \varepsilon \ll 1$,

and

$$\left| \frac{\langle \psi_1(t) | \frac{dH}{dt} | \psi_0(t) \rangle}{[E_1(t) - E_0(t)]^2} \right| < \varepsilon, t \in [0, T]. \tag{3.8}$$

where $\varepsilon \ll 1$.

The above conditions for adiabatic computation are based on Folk theorem, which has come under criticism recently [44]. These conditions do not guarantee general applicability of Adiabatic approximation. A number of attempts have been made to rigorously study Adiabatic Theorem and provide new conditions for generic applicability of adiabatic approximation [45]. Tong et al. have also proposed a set of conditions, which are sufficient for verifying the validity of adiabatic approximation [46]. These are three conditions:

I. $$\left| \frac{\langle \psi_n(t) | \psi_m'(t) \rangle}{E_n(t) - E_m(t)} \right| \ll 1, t \in [0, T] \tag{3.9}$$

II. $$\int_0^T \left| \left(\frac{\langle \psi_n(t) | \psi_m'(t) \rangle}{E_n(t) - E_m(t)} \right)' \right| dt \ll 1 \tag{3.10}$$

III. $$\int_0^T \left| \left(\frac{\langle \psi_n(t) | \psi_m'(t) \rangle}{E_n(t) - E_m(t)} \right)' \right| \left| \langle \psi_n(t) | \psi_l'(t) \rangle \right| dt \ll 1 \tag{3.11}$$

where $m \ne l$, $n \ne m$, T is total evolution time, $|\psi_k(t)\rangle$ and $E_k(t)$ are eigenvectors and corresponding eigenvalues of H at time t, k = 1,. . .,N (N is the size of state space of the system, which includes l, m and n). ψ'_m is first derivative of ψ_m.

A rigorous study has been made of applicability of adiabatic approximation on the basis of these three conditions in case of Grover's Quantum Search under AQC framework in [46]. It has been found that the results obtained originally by Folk theorem is still valid. As this chapter, primarily uses Grover's search, so for rest of the chapter, the results of Folk theorem will be applied.

Grover's Quantum Search algorithm [47] and its variants can be used as a subroutine in designing algorithms for solving optimization problems. The variants of Grover's search algorithm, which do not require prior knowledge of the number of target entities like Boyer's proposal [48] and fixed point schemes [49], are especially suited for such problems. Durr and Høyer proposed an algorithm in [50] for finding minimum value in a database by employing the algorithm proposed in [48] as a subroutine.

The quantum algorithm for locating minima searches for the minimum entry in an un-arranged table 'D' of K items, where every item has a value belonging to an ordered set. The pseudo code for quantum minimum searching algorithm is as follows:

i. Randomly select a threshold index $J_{in} \in [0,K-1]$.
ii. Repeat till the total running time is more than $22.5\sqrt{K} + 1.4 \lg^2 K$.
   a. Apply the Boyer et al. quantum searching algorithm [48].
   b. Observe the first register: let $D[J_{in}']$ be the outcome.
   c. If $D[J_{in}'] < D[J_{in}]$, then set threshold index J_{in} to J_{in}'.
iii. Return J_{in}.

The quantum minimum searching algorithm finds the index of the minimum value and it forms the inspiration for further developing optimization algorithms and heuristics, so in a way, was a path breaking research.

3.3.1 Oracle

A classical oracle of a function f(x) can be implemented reversibly as shown in the Figure 3.1. Similarly regular quantum oracle can be implemented as shown in Figure 3.2 [51]. However, it differs from the reversible oracle as it may be given input in some superposition of basis states and produces output that may be entangled and in some superposition of basis states.

3.4 Grover's adaptive search

Baritompa et al. generalized the quantum minimum search algorithm for solving finite global optimization problems, which have been formulated as follows [52]:

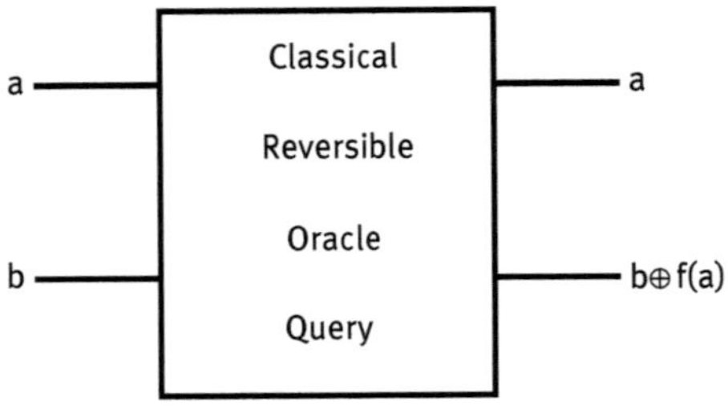

Figure 3.1: Reversible implementation of Classical Oracle.

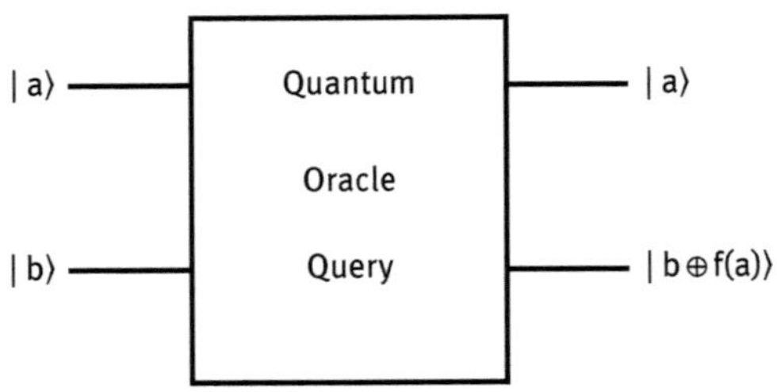

Figure 3.2: Quantum Oracle.

$$\text{Optimize } f(\alpha), \alpha \in A \tag{3.12}$$

where f is a real valued function on a finite set A.

Let the cardinality of set A be N. Let $a_1 < a_2 < a_3 \ldots < a_L$ be 'L' distinct values of the objective function f(α) for all α in A. Therefore, L should be less than or equal to N i.e. if L would be equal to N then for every value of α in set A, there would be a distinct objective function value of f. Let there be a uniform probability measure λ on A, such that ϒ be the range measure given by a stochastic vector $(\Upsilon_1, \Upsilon_2, \ldots, \Upsilon_L)$ induced by f, so that $\Upsilon_j = |f^{-1}(a_j)|/N$ for j = 1, 2, …, L. Let $P_j = \sum_{i=0}^{j} \Upsilon_j$, the probability that a random point has value of 'a_j' or less, i.e. $P_L = 1$. Therefore, there is an improving region corresponding to all values of objective function f(α), which belongs to a sub-set of A, where f(α) has strictly better value [52]. The measure of improving region under λ is denoted as improving fraction P. So if we choose a particular value for objective function, say 'θ', then the improving region would comprise of set of all the point whose objective function values are better than 'θ'.

The algorithm proposed by Baritompa et al. is known as Grover Adaptive search (GAS), which takes a sequence of rotation counts and an objective function value threshold as the input. Its pseudo code is as follows [52]:

i. Select α_1 in A, and set $\theta_1 = f(\alpha_1)$.
ii. Set count nc = 1.

iii. While (termination condition is not met)

{

a. Perform a Grover search of r_{nc} rotations on f with threshold V_{nc}, and denote the outputs by α and θ.
b. If $v < V_{nc}$, set $\theta_{nc+1} = \alpha$ and $V_{nc+1} = v$; otherwise, set $\theta_{nc+1} = \theta_{nc}$ and $V_{nc+1} = V_{nc}$.
c. nc = nc + 1.

}

GAS has been studied using an adaptive search framework developed in [53–55] which has been applied for theoretically analyses of convergence properties of optimization methods that use stochastic global search. One of the underlying assumptions of the adaptive algorithms is that search can find better ("improving") solutions (at some cost). GAS is based on an oracular function that can find better solutions than the current one with a finite probability in every iteration, if one exists. This oracle takes an input string $\alpha \in A$, computes f(α), compares f(α) with current threshold and outputs zero or one depending on the result of comparison and the type of optimization problem being solved i.e. whether it is a minimization or maximization problem. Such oracles can be implemented using reversible / quantum circuits as shown in Section 3.3.1.

Thus, GAS generates a series of domain solutions, which are uniformly distributed in the improving region of the previous solution. Further, according Theorem 1 and Corollary 6 given in [54], such a series converges to a global optimal solution rapidly; e.g., a unique optimal solution in a domain of size K can be located after $1 + \ln(K)$ such improvements, in expectation [54].

Theorem 1 [54]. *The expected number of iterations, I, required for solving the finite optimization problem described by* eq. (3.12) *is:*

(a) $I = 1 + \sum_{i=2}^{N} \frac{\Upsilon_i}{P_i}$ *for strong Pure Adaptive Search.*

(b) $I = 1 + \sum_{i=2}^{N} \frac{\Upsilon_i}{P_{i-1}}$ *for weak Pure Adaptive Search.*

where Strong Pure Adaptive Search implies that each iteration of the search algorithm should sample objective function value from a strictly improving region. The weak Adaptive search implies that each iteration of the search algorithm should sample objective function value that is either equal or better.

Corollary 6 [54]. *The expected number of iterations, I, for solving finite global optimization described by* eq. (3.12), *and given a uniform distribution on the objective function values, is:*

(a) *bounded above by 1 + logN for strong Pure Adaptive Search.*
(b) *bounded above by 2 + log(N–1) for weak Pure Adaptive Search.*

GAS corresponds to the weaker version of Pure Adaptive search as it cannot be guaranteed that after application of Grover Search, an strictly better solution would be found. It can be argued that Grover Search can return an objective function value, which is less than the threshold and so it does not corresponds to Pure Adaptive Search Framework, but, in such a case, the inferior value would be discarded and the current threshold would be treated as the output of the algorithm for the particular iteration. Thus, the overall GAS corresponds to weaker version of Pure Adaptive Search.

An important design decision of Grover's Adaptive Search is determination of the rotation count sequence. It is possible to determine the optimal rotation count 'r' for every iteration of the GAS provided the information regarding improving fraction 'P' of the domain is available. Thus, if the optimal rotation count is used for every run of Grover's Search, then a better (improved) solution will almost certainly be found in every iteration of the GAS. However, in real world problems, the improving fraction *P* is often unknown. An estimate of the improving fraction P can always be made by using all the information available at the end of an iteration and considering the nature of the problem. That is a Bayesian approach can be applied by tracking the improving fraction at every iteration through the sequence of posterior distributions, which will be subsequently used for choosing rotation count at the particular iteration. However, this kind of approach might be very complicated and cumbersome [56]. Thus, other methods have been devised to strike a balance between ease of implementation and optimality of rotation count selection, which includes method based on Boyer et al. [48] used by Durr and Hoyer illustrated in [50].

Another method proposed by Baritompa et al. is based on maximizing the benefit to cost ratio denoted by parameter b in the heuristic [52]. The benefit has been taken to be the expected decrease in the improving fraction of the domain, which is difficult to ascertain without knowing the nature of the problem being solved. The cost is dependent on the rotation count. The pseudo code for heuristic is as follows:

I. Initialize $u_0(\Omega) = \Omega$ // u_0 is a polynomial
II. For $i = 1, 2, \ldots,$ do: // Generating rotation count r for ith iteration

A. $E_{ui} = 1 - \int_0^1 u_i(\Omega)d\Omega$
B. $\beta = 0$
C. For $r = 0, 1, \ldots,$ until $E_{ui}/(r+1) < 2\beta$, do:

(i) $v(\Omega) = u_i(\Omega) + \Omega \int_\Omega^1 \frac{g_r(P)}{P} du_i(P)$ $//g_r(P) = sin^2((2r+1)sin^{-1}\sqrt{P})$

(ii) $E_v = 1 - \int_0^1 v(\Omega)d\Omega$

(iii) $b = \frac{(E_{ui} - E_v)}{r+1}$

(iv) If $b > \beta$, then

a) $\hat{r} = r$

b) $\beta = b$

c) $u_{i+1}(\Omega) = v(\Omega)$

D. Output $\hat{r}$

The sequence of rotation counts produced by the heuristic is independent of the particular optimization problem and is as follows for first 33 entries [52]:

0, 0, 0, 1, 1, 0, 1, 1, 2, 1, 2, 3, 1, 4, 5, 1, 6, 2, 7, 9, 11, 13, 16, 5, 20, 24, 28, 34, 2, 41, 49, 4, 60, ...

Recently, the above heuristic method has been further studied and improved by Liu and Koehler [56]. One improvement has been to simplify the calculations involved in computing distributions by employing Chebyshev polynomials, which considerably reduced the computation time. The other improvement has been to convert the above heuristic into a dynamic one, which takes into account whether the Grover search in a particular iteration was successful or not. This has been made possible by using Bayes Law to update the distribution after one complete round of Grover Search depending on success or failure of finding a better solution. The heuristic for computing dynamic rotation schedule is as follows:

I. Initialize $u_0(\Omega) = \Omega$

II. For $i = 1, 2, \ldots,$ do: // Generating rotation count r for ith iteration

A. $E_{ui} = 1 - \int_0^1 u_i(\Omega) d\Omega$

B. $\beta = 0$

C. For $r = 0, 1, \ldots,$ until $E_{ui}/(r+1) < 2\beta$, do:

(i) $b = \int_0^1 P g_r(P) du_i(P)$

(ii) If $b > \beta$, then

a) $\hat{r} = r$

b) $\beta = b$

D. Output $\hat{r}$

E. If (Search_outcome = Success)

$$u_{i+1}(\Omega) = \frac{\int_0^y g_r(P)du_i(P) + \int_y^1 \frac{y}{P} g_r(P)du_i(P)}{\int_0^1 g_r(P)du_c(P)}$$

Else

$$u_{i+1}(y\Omega) = \frac{u_i(\Omega) - \int_0^y g_r(P)du_i(P)}{1 - \int_0^1 g_r(P)du_c(P)}$$

Liu and Koehler have reported that formal derivation of the expected time required by the proposals, made by Durr and Hoyer [50] (based on Boyer et al. algorithm [48]), Baritompa et al. [52] and their own improved and dynamic version [56] of quantum algorithms for global optimization, is hard to derive. This is because the basic subroutine employed for searching needs knowledge of the distribution function of the improving fraction. Therefore, numerical simulations have been used to compare the methods. However, such simulation studies are computationally very expensive as they require digital computer to simulate a quantum computer. Thus, the performance of such algorithms could be investigated on only problems of small size, which have not much practical significance.

This chapter identifies that basic problem in the above global optimization algorithms is due to the dependence of basic search subroutine on the improving fraction to determine the rotation count in a particular iteration. This problem can be solved by changing quantum computation model from quantum circuit model to Adiabatic Quantum Computation (AQC) model.

AQC provides a distinct advantage over quantum circuit model on Grover's Search i.e. it eliminates the problem of overshooting. This is possible because the solution is encoded as the ground state of the Problem Hamiltonian, therefore, even if the system is evolved for time period longer than required, the system remains in the ground state of the problem Hamiltonian. This feature of AQC makes it possible to formally derive the expected running time of Global Optimization Algorithm based on AQC.

3.5 Problem formulation

Optimization problems that are regularly solved in machine learning and engineering domain are mostly in continuous domain. Such problems are classified as unconstrained and constrained global optimization problems. The unconstrained global optimization problems (UCOP) in continuous variables are generally formulated as follows:

Optimize $f(x)$ where $x = (x_1, x_2, \ldots.., x_E)\ \varepsilon\ R^E$,

Such that:

$$x_{il} < x_i < x_{iu}; \text{where } x_i \text{ is the } i^{th} \text{ variable with } x_{il} \text{ and } x_{iu} \text{ as its lower and upper limits.} \tag{3.13}$$

The objective function $f(x)$ as well as the inequality and the equality constraints are often nonlinear, non-convex, non-differentiable, multimodal and of high dimension. Thus, optimization techniques, which are efficient by being deterministic, like gradient based methods and exhaustive search, are often rendered ineffective for solving such ill-defined problems [57]. Calculus based methods typically perform local optimization by using search space domain information like local gradient [58]. Enumerative Strategies suffer from the curse of dimensionality i.e. the runtime complexity increases exponential with the dimension of the problem [59].

It has been shown in Section 3.4 that Quantum global optimization algorithms can be used for solving finite search domain problems. The search domain of any real variables problem is infinite. In order to overcome this limitation of quantum algorithms, the real variables search space in [60] has been discretized. That is a variable is represented by a finite number of qubits, which determines the resolution and the level of discretization. For example, the decision variables, x_i, in continuous domain are bounded by x_{il} and x_{iu}. The search space for variable x_i can be discretized by taking n_i qubits, so that there will be 2^{ni} distinct values of x_i which can be selected between the range of $[x_{il}, x_{iu}]$. The maximum discretization error, Δe_i, is as follows:

$$\Delta e_i = (x_{iu} - x_{il})/2^{(ni+1)} \tag{3.14}$$

This discretization error is present even in computation performed with classical digital computers. The total number of qubits, nq, required to map the discretized solution space is as follows:

$$nq = \sum_i n_i \tag{3.15}$$

The larger the number of qubits more is the accuracy, however, it also increases the running time of the quantum algorithm. Therefore, there is a trade-off involved between accuracy and computation time. An output or auxiliary qubit is also required which evolves according to the result of the computation of the objective function implemented by a reversible circuit [52].

3.6 AQC based quantum search

Adiabatic Quantum Computation is analog analogue of discrete gate-based quantum computing, which is a generic framework that depends on Hamiltonian and Adiabatic theorem [31]. A Problem Hamiltonian (H_{fin}) is designed for the problem being solved such that it is a time independent Hamiltonian, and the ground state of H_{fin}, $|\psi_{fin}\rangle$, represents a solution of the problem. The quantum system is evolved by a time dependent Hamiltonian, H(t). H(0) corresponds to a time independent Hamiltonian known as initial Hamiltonian (H_{ini}), whose ground state is easy to construct. Hamiltonian H(t) is time-varying and governs the evolution of the system and it varies according to the bounds given by adiabatic theorem, from H_{ini} to H_{fin}, so that the quantum system stays near the ground state of the instantaneous Hamiltonian H(t) till it reaches H_{fin}. Thus, Adiabatic Quantum Algorithm has following main steps:

1) Determine the number of qubits, nq, required to represent the problem in the quantum system.
2) Find an easily constructible initial ground state and construct its corresponding H_{ini}.
3) Design the H_{fin} according to the problem being solved.
4) Construct the time dependent Hamiltonian H(t), which will govern the evolution of the system for time T:

$$H(t) = H_{ini}{}^*(1-t/T) + H_{fin}{}^*(t/T),$$

   or,

$$\hat{H}(s) = H_{ini}{}^*(1-s) + H_{fin}{}^*(s), \tag{3.16}$$

   where s = t/T and s(0) = 0 and s(T) = 1.
5) Estimate total evolution time T.
6) Evolve the Quantum System according to Schrodinger's Equation for time T:

$$i\hbar\frac{d|\psi(t)\rangle}{dt} = H(t)|\psi(t)\rangle \tag{3.17}$$

   where ħ = 1.
7) Perform Measurement on the Quantum system after time T to obtain the solution of the problem.

Search for a marked element in an unstructured database with N_e elements was also attempted by Farhi et al. in [31] by using global version of the Adiabatic Theorem for their quantum algorithm. This resulted in a running time of the order of N i.e. O(N), which was no better than a classical algorithm. Roland and Cerf improved the performance of Adiabatic Quantum Algorithm for this problem by using local adiabatic evolution approach [32]. Thus, Adiabatic version of Grover's search also has the same complexity as that of Circuit model. This chapter employs local version of Adiabatic evolution proposed by Roland and Cerf in [32] for Grover's Search to design a quantum algorithm for solving optimization problems in continuous domain.

The algorithm starts with initial state as $|\varphi_O\rangle$, which is in equal superposition of all the basis states, N_e (j represents index of elements in unsorted database):

$$|\phi_O\rangle = \frac{1}{\sqrt{N_e}} \sum_{j=0}^{Ne-1} |j\rangle \tag{3.18}$$

The beginning Hamiltonian, H_{ini}, of the system is chosen as:

$$H_{ini} = I - |\varphi_O\rangle\langle\varphi_O|, \tag{3.19}$$

where I is the Identity matrix and $|\varphi_O\rangle$ is its ground state with zero energy. Therefore, the H_{ini}, for quantum search algorithm is independent of the problem. Let us assume that there exist a problem Hamiltonian H_{fin}, which can be represented as follows:

$$H_{fin} = I - \sum_{M \in m_e} |\phi_M\rangle\langle\phi_M|, \tag{3.20}$$

where m_e is the ensemble of solutions and $|\varphi_M\rangle$ represents the ground state, and is a solution. H_{fin} can be applied without explicitly knowing $|\varphi_M\rangle$, by using an oracle to tell whether the solution has been found or not. It is implemented as follows:

$$H_{fin}|\varphi_i\rangle = \text{fpb}(\varphi_i)|\varphi_i\rangle \tag{3.21}$$

where function $\text{fpb}(\varphi_i)$ has been implemented as follows:

$$\text{If}(|\varphi_i\rangle \in \{|\varphi_M\rangle\})$$

$$\text{fpb}(\varphi_i) = 0$$

else

$$\text{fpb}(\varphi_i) = 1$$

That is, if $|\varphi_i\rangle$ is the marked state, then $\text{fpb}(\varphi_i)$ is equal to zero, corresponding to the ground state, i.e. lowest energy and if $|\varphi_i\rangle$ is not the marked state, then $\text{fpb}(\varphi_i)$ is equal to one, i.e. having higher energy level than the ground state. The $\text{fpb}(\varphi_i)$ can be reversibly implemented by using the same technique as described for oracles in Section 3.3.1.

The time dependent Hamiltonian that evolves the system under initial Hamiltonian to final Hamiltonian is as follows:

$$H(t) = (1 - s(t))\, H_{ini} + s(t)H_{fin} \tag{3.22}$$

where s(t) is a continuous function with s(0) = 0 and s(T) = 1. Let eigenstates of H(t) be $|\psi_k(t)\rangle$ such that:

$$\begin{aligned} \mathrm{fpb(q)} &= 0; \\ \mathrm{fpb(q)} &= 1; \end{aligned} \tag{3.23}$$

where $E_k(t)$ are the corresponding eigenvalues and k labels the eigenstates (k = 0 indicates the ground state).

The gap between the lowest two eigenvalues, E_0 and E_1, is described by the following equation [32]:

$$g(s) = \sqrt{1 - 4\left(1 - \frac{m_e}{N_e}\right) s(t)(1 - s(t))} \tag{3.24}$$

$$g_{\min} = \sqrt{}(m_e/N_e) \text{ and occurs at } s = 1/2.$$

The matrix element of dH/dt in eq. (3.5) can be alternately expressed as follows:

$$\left\langle \frac{dH}{dt} \right\rangle_{1,0} = \frac{ds}{dt} \left\langle \frac{d\tilde{H}}{ds} \right\rangle_{1,0} \tag{3.25}$$

Using eqs. (3.5) and (3.25), for all times t, we get new condition

$$\left|\frac{ds}{dt}\right| \leq \varepsilon \frac{g^2(s)}{\left|\left\langle \frac{d\tilde{H}}{ds} \right\rangle_{1,0}\right|} \tag{3.26}$$

Using eq. (3.26) and bounding $\left|\left\langle \frac{d\tilde{H}}{ds} \right\rangle_{1,0}\right| \leq 1$, the Hamiltonian H(t) is evolved at a following rate:

$$\frac{ds}{dt} = \varepsilon g^2(s) = \varepsilon \left\{ 1 - 4\left(1 - \frac{m_e}{N_e}\right) s(1 - s) \right\} \tag{3.27}$$

After integration, we get:

$$t = \frac{1}{2\varepsilon} \left\{ \frac{N_e}{\sqrt{m_e(N_e - m_e}} \right\} \left[\tan^{-1}\left\{ \left(\sqrt{\frac{N_e - m_e}{m_e}} \right) (2s - 1) \right\} + \tan^{-1}\left(\sqrt{\frac{N_e -}{m_e}} \right) \right] \tag{3.28}$$

And from the above eq. (3.28), s(t) can be obtained as follows:

$$s(t) = \frac{1}{2} \left\{ \left(\sqrt{\frac{m_e}{N_e - m_e}} \right) \tan\left[2\varepsilon \left\{ \frac{\sqrt{m_e(N_e - m_e)}}{N_e} \right\} t - \tan^{-1}\left(\sqrt{\frac{N_e - m_e}{m_e}} \right) \right] + 1 \right\} \tag{3.29}$$

Therefore, the total computation time T can be calculated by equating s = 1 in eq. (3.28), we get:

$$T = \frac{1}{\varepsilon}\left\{\frac{N_e}{\sqrt{m_e(N_e - m_e)}}\right\}\tan^{-1}\left(\sqrt{\frac{N_e - m_e}{m_e}}\right) \tag{3.30}$$

$$\max\left\{\tan^{-1}\left(\sqrt{\frac{N_e - m_e}{m_e}}\right)\right\} \le \frac{\pi}{2}\ \forall\ \mathrm{N_e} > m_e \tag{3.31}$$

$$\left\{\frac{N_e}{\sqrt{m_e(N_e - m_e)}}\right\} \le \left\{\sqrt{2}\left(\sqrt{\frac{N_e}{m_e}}\right)\right\}\ \forall\ m_e \le \frac{N_e}{2} \tag{3.32}$$

Using above three equations i.e. from (3.30) to (3.32), we can conclude that

$$T \le \frac{\pi\sqrt{2}}{2\varepsilon}\left(\sqrt{\frac{N_e}{m_e}}\right)\ \forall\ m_e \le \frac{N_e}{2} \tag{3.33}$$

The above result is very useful from the point of quantum search based on AQC for solving optimization problems, which have been attempted in [52]. Further, the result in [32] was derived for cases in which $m_e \ll N_e$. Thus, the bounds provided here are applicable to more generic cases.

One observation, which shows that AQC is a better model than Quantum Circuit model for implementing canonical Grover's search is that, it does not overshoot the marked element if the system is evolved for time longer than required i.e. if the rate of evolution given by eq. (3.27) is slower than necessary, there is no overshooting involved as the system remains in the eigenstate of the final Hamiltonian. That is if the system is evolved at $g_{min} = 1/\sqrt{N_e}$ instead of $g_{min} = \sqrt{(m_e/N_e)}$, where $m_e > 1$, then it just takes more time to evolve than necessary, but the solution is still reached, unlike Grover's search on circuit model in which if more rotations than required are given to the state vector, then it overshoots the target state. However, in such cases, the overall computational complexity is $O(\sqrt{N_e})$ instead of $O(\sqrt{(N_e/m_e)})$, which when m_e is large would be a substantial slow down.

3.7 AQC based minimum searching algorithm

The algorithm presented by Durr and Hoyer [50] on quantum circuit model can be modified to fit into AQC framework as follows:

a. Set $m_e = N_e/2$ and termination counter I = 0.
b. Select the initial threshold, Yth, by randomly taking a sample.
c. Initialize H_{ini} as given in eq. (3.19).

d. Let H_{fin} be as given by eq. (3.20).

where function fpb (φ_i) has been implemented as follows:

$$\text{If}(\text{D}[\varphi_{\text{i}}] < \text{Yth}) \ // \ \text{D is the table}$$

$$\text{fpb}(\varphi_{\text{i}}) = 0$$

else

$$\text{fpb}(\varphi_{\text{i}}) = 1$$

e. Employ Hamiltonian $\hat{H}(s) = (1\text{-}s(t))^{*}H_{ini} + s(t)^{*}H_{fin}$, where s(t) is as given eq. (3.29).
f. Perform evolution of the system for time T as per eq. (3.1). The time T is given by eq. (3.33).
g. Measure the qubits in $|\Psi(T)\rangle$ to find the solution of the problem and present result by classical evaluation as objective function value, y.
h. If D[y] is less than threshold Yth, then

$$\text{Yth} = \text{D}[\text{y}]$$

i. If $I > T_{tot}$, then Stop
j. $I = I + T$
k. $m_e = m_e - 1$
l. Go to c.

Theorem 1: The above algorithm finds the minimum in a database in $O(\sqrt{N_e})$ time with error probability ε^2 in expectation.

Proof: Let us rank all the elements in the database with rank 1 assigned to the element with minimum value and rank N_e assigned to the element with maximum value. All other elements are also ranked according to their position, if there are several elements with same value, then rank them sequentially as they appear without breaking order with elements having different values.

The above algorithm would move through a succession of points before finding the minima, therefore, assume that it starts with the element as threshold, whose rank is $N_e/2$, (which can be found by either randomly choosing the threshold by sampling the database a few times) then the total running time, T_{tot}, of the algorithm is as follows;

$$T_{tot} = \sum_{r=1}^{N/2} \{P(N_e, r_n) T(N_e, r_n - 1)\} \tag{3.34}$$

where $P(N_e,r_n)$ is the probability of the element with rank r_n ever being selected and $T(N_e,r_n-1)$ is the expected time required to evolve the system for finding an element with a better rank than r_n.

The $P(N_e,r_n)$ has been shown to be independent of the database size in lemma 1 of [50] and is dependent only on its rank r_n.

$$P(N_e, r_n) = 1/r_n \tag{3.35}$$

The $T(N_e, r_n-1)$ can be found by inserting the value of r_n in eq. (3.33):

$$T(N_e, r_n - 1) = \frac{\pi\sqrt{2}}{2\varepsilon}\left(\sqrt{\frac{N_e}{r_n - 1}}\right) \tag{3.36}$$

Therefore,

$$T_{tot} = \sum_{r_n=2}^{\frac{N_e}{2}}\left[\frac{1}{r_n}\left\{\frac{\pi\sqrt{2}}{2\varepsilon}\left(\sqrt{\frac{N_e}{r_n-1}}\right)\right\}\right] = \left(\frac{\pi\sqrt{2}}{2\varepsilon}\right)\sqrt{N_e}\sum_{r_n=1}^{\frac{N_e}{2}}\left[\left(\frac{1}{r_n+1}\right)\left(\frac{1}{\sqrt{r_n}}\right)\right] \tag{3.37}$$

$$T_{tot} = \left(\frac{5\pi\sqrt{2}}{4\varepsilon}\right)\sqrt{N_e} \tag{3.38}$$

Thus, the AQC based minimum finding algorithm can be analyzed analytically and has a better running time in expectation.

3.8 Adiabatic quantum optimization algorithm

The Adiabatic quantum optimization algorithm has been developed for solving unconstrained optimization problems, which have been described in Section 3.5. The AQC based minimum finding algorithm has been used as the underlying framework for developing the optimization algorithm.

The unconstrained optimization problems in continuous domain can be converted into a finite global optimization problem as shown in Section 3.5. Each domain point can be ranked on the basis of its objective function value i.e. the point with best objective function value can be assigned RANK 1 and the point with worst objective function value can be assigned RANK N_e. The rest of points can similarly be assigned a suitable rank. If two or more points have same objective function value then they would be assigned different ranks, randomly chosen within the range, while maintaining their relative position with rest of the points. Thus, the points are now analogues to the index of the database, so, the optimization algorithm can be designed on the same framework as the minimum finding algorithm described in Section 3.5.

Adiabatic quantum optimization algorithm for solving Unconstrained Global Optimization problem is as follows:

a. Set $m_e = N_e/2$ and termination counter $I = 0$.
b. Compute Yth by taking a random sample of domain point x.

c. Construct Beginning Hamiltonian H_{ini}, which would be same for all the UCOP problems.

$H_{ini} = \sum H_{ini\_i}$, where

$H_{ini\ i} = I - |\psi_i\rangle\langle\psi_i|$, where

$|\psi_i\rangle = (1/\sqrt{2})^*(0\rangle + (1/\sqrt{2}))$

d. Make Problem Hamiltonian H_{fin}, which is specific to the particular UCOP, however the structure would remain same irrespective of the UCOP.

$$H_{fin}|q_1q_2q_3...q_n\rangle = fpb\ (q_1q_2q_3...q_n)\ |q_1q_2q_3...q_n\rangle$$

where $fpb(q_1q_2q_3...q_n)$ is a Boolean function, which models the problem and can be constructed reversibly as follows:
If ($f(q) \geq Yth$) …… for Minimization

$$fpb(q) = 0;$$

else

$$fpb(q) = 1;$$

where $|q_1q_2q_3...q_n\rangle$ are n qubits representing the decision variables of the UCOP being solved given eq. (3.13) and $q = q_1q_2q_3...q_n$. The fpb(q) would be implemented in an Oracle and so its outcome would be reflected in the state of ancilla qubit, qo. Yth is the threshold value computed from the objective function of the UCOP being solved. The first value of Yth may be assigned randomly or seeded by meta-heuristics techniques [57].

e. Employ Hamiltonian $\hat{H}(s) = (1-s(t))^*H_{ini} + s(t)^*H_{fin}$, where s(t) is as given eq. (3.29).

f. Perform evolution of the system for time T as per eq. (3.1). The time T is given by eq. (3.33).

g. Measure the qubits in $|\Psi(T)\rangle$ to find the solution of the problem and present result by classical evaluation as objective function value, y.

h. If y is less than threshold Yth, then

$$Yth = y - \Delta y \text{ where } \Delta y \geq 0 \text{ for minimization problem}$$

i. If $I > T_{tot}$, then Stop, where T_{tot} is given by eq. (3.38)

j. $I = I + T$

k. $m_e = m_e - 1$

l. Go to c.

3.8.1 Discussions

The nucleus of the proposed algorithm is a search routine designed on AQC framework to find optimal value. It adapts the evolution rate to local adiabatic condition based on local adiabatic evolution, which has $O(\sqrt{N_e})$ run-time complexity and is asymptotically optimal as shown in [32].

The termination of the run of algorithm for solving a particular instance of the problem is also supported by the laws of Quantum mechanics (i.e. Adiabatic Theorem) that is if no improvement is made even after considering $m_e = 1$, then the search can be terminated with conclusion that no better points exists with probability $1\text{-}\varepsilon^2$, where $\varepsilon \ll 1$. This consideration makes the proposed algorithm a global optimizer.

The working of the algorithm is further illustrated with the help of an example. Let us Minimize $f(x) = x^2$ such that $-1.0 < x < 1.0$

The variable x is discretized between [−1.0, 1.0] by using 10 qubits i.e. $n = 10$.

The total number of domain points in search space, $N = 2^n = 2^{10} = 1024$.

For first iteration: Let $\varepsilon = 0.707$, $x = 0.5$; $Y_{th} = (0.5)^2 = 0.25$, $m_e = 512$, therefore $T = 2.22$ time units. If the run is successful, a new threshold would be found or else with the earlier threshold, and after reducing m_e by 1, the second iteration would be executed, so on and so forth. Subsequently, say, after 256 iterations, lf $|x| > 0.25$, then it would have a high chance of success run as T for which it would be evolved is larger than required, whereas if $|x| < 0.25$, then T is less than required but in next few iterations, the T would become equal or larger (as m_e is reduced) than required for a success run. Moreover, when $|x| < 0.25$, the performance of the algorithm is better than expected. Therefore, if the algorithm fails, it is because its performance is better than expected otherwise it is almost guaranteed to produce a good result. Thus, after running the algorithm for T_{tot} time, it can be re-run for $m_e = 1$ and if no improvement is found in threshold then it can be said that global optimal has been found with failure probability at most ε^2.

The actual running time of the algorithm can be improved further by doing some additional processing like counting the successful and unsuccessful iterations and for updating the value of m_e as the eq. (3.38) gives an average case result. However, the analytical analysis of such adaptive schemes would be difficult and is left for future studies.

AQC optimization algorithm is better than the Quantum circuit implementation of Grover Adaptive Search as the rotation count is no longer required to be guessed or randomly / heuristically assigned. Further, even if the quantum system were evolved for time duration larger than the minimum time, the system would only get closer to the ground state and would not overshoot the marked state containing the solution.

3.9 Conclusions

Machine learning is a branch of Artificial intelligence, which tries to develop techniques that have ability to learn from experiences. These experiences are articulated as data generated from underlying models. The goal of any machine learning technique is to abstract the underlying model reliably for specific tasks like prediction and decision making. Optimization problems are ubiquitous in machine learning and other domains like engineering and a number of specialized techniques for efficiently solving specific problems have been developed. However, general purpose techniques like enumerative strategies have run time complexity of O(N), where N is the number of domain points. Grover's quantum search algorithm provides quadratic speed up over its classical counter parts. Baritompa et al. integrated ideas from Grover's search into the framework of Pure adaptive search and called the new framework as Grover Adaptive search. It was further improved by Liu and Koehler. However, GAS was essentially an optimization heuristic due to which analytical analysis was difficult.

A novel quantum optimization algorithm based on Adiabatic Quantum Computation framework, which employs local evolution instead of global evolution, has been proposed. It is also inspired by the application of Grover's Search algorithm for finding minimum solution in a database, however, the proposed algorithm solves real domain optimization problems, which are routinely solved in Machine learning algorithms. Further, the proposed algorithm has been analytically analyzed and has been shown to have the same signature of quadratic speed up of Grover's search algorithm. Baritompa et al. had made a similar effort on gate-based quantum computation model for solving unconstrained global optimization problem. However, the proposed AQC based algorithm is conceptually better than gate-based model as there is no fear of overshooting the optima, which is a major concern in gate-based models, when a priori knowledge of the number of optimal solutions is not available, whereas the optimal solutions are encoded as the ground state of the problem Hamiltonian. The problem of overshooting in gate-based model could further increase convergence time in real world implementations.

In future, we would like to develop a simulator and architecture of a hybrid Classical and Adiabatic Quantum Computer on which we can demonstrate operations of our proposal of Quantum Optimization Algorithm for Machine Learning techniques, which would require addressing several challenging issues like implementation of loss function on a quantum computer and weights updating in learning mechanism.

References

[1] Kline, R. R. Cybernetics, Automata Studies and the Dartmouth Conference on Artificial Intelligence. IEEE Annals of the History of Computing, October–December, 2011, IEEE Computer Society.

[2] Russell, S., & Norvig, P. Artificial Intelligence: A Modern Approach, 3rd ed. Prentice Hall Press Upper Saddle River, NJ, USA, 2009.
[3] Columbus, L. Why AI Is The Future of Financial Services. August, 2019. https://www.forbes.com/sites/louiscolumbus/2019/08/15/why-ai-is-the-future-of-financial-services/#452381ba3847
[4] Insight Team. AI And Healthcare: A Giant Opportunity. Februrary, 2019. https://www.forbes.com/sites/insights-intelai/2019/02/11/ai-and-healthcare-a-giant-opportunity/#35e679924c68
[5] Grühn, B., Kässer, M., Köstring, J., Padhi, A., & Tschiesner, A. Winning tomorrow's car buyers using artificial intelligence in marketing and sales. February, 2019. https://www.mckinsey.com/industries/automotive-and-assembly/our-insights/winning-tomorrows-car-buyers-using-artificial-intelligence-in-marketing-and-sales
[6] Teich, D. A. Artificial Intelligence Can Address Social Media Risks And Rewards. March, 2019. https://www.forbes.com/sites/davidteich/2019/03/27/artificial-intelligence-can-address-social-media-risks-and-rewards/#295c4dc9686c
[7] Hoy, M. B. Alexa, Siri, Cortana, and More: An Introduction to Voice Assistants. Medical Reference Services Quarterly. 2018, 37(1),81–88. doi:10.1080/02763869.2018.1404391. PMID 29327988
[8] Zhang, G., Patuwo, B. E., & Hu, M. Y. Forecasting with artificial neural networks: The state of the art. International Journal of Forecasting, 1998, 14(1),35–62, ISSN 0169-2070. https://doi.org/10.1016/S0169-2070(97)00044-7
[9] Chang, C., & Ling, C. LIBSVM: A library for support vector machines. ACM Transactions on Intelligent Systems and Technology. 2011, 2(3). 10.1145/1961189.1961199
[10] Jain, A. K. Data clustering: 50 years beyond K-means. Pattern Recognition Letters. 2010, 31 (8),651–666, ISSN 0167-8655. https://doi.org/10.1016/j.patrec.2009.09.011
[11] Liu, H., Miao, E. M., Wei, X. Y., & Zhuang, X. D. Robust modeling method for thermal error of CNC machine tools based on ridge regression algorithm. International Journal of Machine Tools and Manufacture. 2017, 113, 35–48, ISSN 0890-6955. https://doi.org/10.1016/j.ijmachtools.2016.11.001
[12] Garg, P., Neider, D., Madhusudan, P., & Roth, D. Learning invariants using decision trees and implication counterexamples. Proceedings of the 43rd Annual ACM SIGPLAN-SIGACT Symposium on Principles of Programming Languages (POPL '16). ACM, New York, NY, USA, 2016, 499–512. https://doi.org/10.1145/2837614.2837664
[13] Tangirala, A. K. Principles of System Identification: Theory and Practice 1st Edition. CRC Press, 2014.
[14] LeCun, Y., Bengio, Y., & and Hinton, G. Deep learning. Nature. 2015, 521, 436–444.
[15] Poplin, R., Varadarajan, A. V., Blumer, K., Liu, Y., McConnell, M. V., Corrado, G. S., Peng, L., & Webster, D. R. Prediction of cardiovascular risk factors from retinal fundus photographs via deep learning. Nature Biomedical Engineering. 2018, 2, 158–164.
[16] Rebentrost, P., Mohseni, M., & Lloyd, S. Quantum support vector machine for big data classification. Phys. Rev. Lett. 2014, 113, 130503.
[17] Giovannetti, V., Lloyd, S., & Maccone, L. Quantum Random Access Memory. Phys. Rev. Lett. 2008, 100, 160501.
[18] Optimization for Machine Learning. Sra, S., Nowozin, S., & Wright, S. J. (Eds). The MIT Press, 2011.
[19] Rebentrost, P., Schuld, M., Wossnig, L., Petruccione, F., & Lloyd, S. Quantum gradient descent and Newton's method for constrained polynomial optimization. New J. Phys. 2019, 21, 073023.

[20] Klymko, C., Sullivan, B. D., & Humble, T. S. Adiabatic quantum programming: minor embedding with hard faults. Quantum Inf Process. 2014, 13(709). https://doi.org/10.1007/s11128-013-0683-9
[21] Harrow, A. W., Hassidim, A., & Lloyd, S. Quantum algorithm for linear systems of equations. Phys. Rev. Lett. 2009, 103, 150502.
[22] Farhi, E., Goldstone, J., Gutmann, S., Lapan, J., Lundgren, A., & Preda, D. A quantum adiabatic evolution algorithm applied to random instances of an NP-complete problem. Science. 2001, 292(5516),472–476.
[23] Farhi, E., Goldstone, J., Gutmann, S., & Nagaj, D. How to Make the Quantum Adiabatic Algorithm Fail. International Journal of Quantum Information. 2008, 6(503).
[24] Reichardt, B. The quantum adiabatic optimization algorithm and local minima. Proceedings of the 36th Annual Symposium on the Theory of Computing (IEEE). Computer Society Press, New York, 2004, 279–287.
[25] van Dam, W., Mosca, M., & Vazirani, U. How powerful is adiabatic quantum computation? Proceedings of 42nd Annual IEEE Symposium on Foundations of Computer Science. 2001, 279–287.
[26] Amin, M. H. S. Effect of Local Minima on Adiabatic Quantum Optimization. Physical Review Letters. 2008, 100, 130503.
[27] Farhi, E., Goldstone, J., Gutmann, S., Meyer, H. B., & Shor, P. Quantum Adiabatic Algorithms, Small Gaps, and Different Paths. 2009. arXiv:0909.4766v2 [quant-ph]
[28] Altshuler, B., Krovi, H., & Roland, J. Anderson localization makes adiabatic quantum optimization fail. Proceedings of National Academy of Science. USA, 2010, 107(28),12446–12450.
[29] Choi, V. Different Adiabatic Quantum Optimization Algorithms for the NP-Complete Exact Cover and 3SAT Problems. 2011. arXiv:1010.1221v3
[30] Dickson, N. G., Amin, M. H. S. Does Adiabatic Quantum Optimization Truly Fail for NP-complete problems? 2011. arXiv:1010.0669
[31] Farhi, E., Goldstone, J., Gutmann, S., & Sipser, M., Quantum Computation by Adiabatic Evolution. 2000. arXiv:quant-ph/0001106v1
[32] Roland, J. & Cerf, N. Quantum search by local adiabatic evolution, Physical Review A. 2002, 65(4), 042308.
[33] Rezkhani, A. T., Pimachev, A. K., & Lidar, D. A. Accuracy versus run time in an adiabatic quantum search. Physical Review A. 2010, 82, 052305.
[34] Zhang, Y.Y & Lu, S.F. Quantum search by partial adiabatic evolution. Phys. Rev. A. 2010, 82, 034304.
[35] Aharonov, D., Dam, W. van Kempe, J., Landau, Z., Lloyd, S., & Regev O. Adiabatic quantum computation is equivalent to standard quantum computation. Proceedings of 45th Annual IEEE Symposium on Foundations of Computer Science. 2004, 42–51.
[36] Mizel, A., Lidar, D. A., & Mitchell, M. W. Simple proof of equivalence between adiabatic quantum computation and the circuit model. Phys. Rev. Lett. 2007, 99, 070502.
[37] Brady, L. T. & van Dam, W. Necessary adiabatic run times in quantum optimization. Phys. Rev. 2017, A 95, 032335.
[38] Guerreschi, G. G. & Matsuura, A. Y. QAOA for Max-Cut requires hundreds of qubits for quantum speed-up. Scientific Reports. 2019, 9, 6903.
[39] Farhi, E., Goldstone, J., & Gutmann, S. A Quantum Approximate Optimization Algorithm. 2014. http://arxiv.org/abs/1411.4028v1
[40] McClean, J. R., Romero, J., Babbush, R., & and Aspuru-Guzik, A. The theory of variational hybrid quantum-classical algorithms. New J. Phys. 2016, 18, 023023.
[41] Mani, A. & and Patvardhan, C. Some Investigations for Solving Constrained Optimization Problems using Adiabatic Quantum Computation. 2011 Eighth International Conference on

Information Technology: New Generations. Las Vegas, NV, 2011, 1101–1102. doi:10.1109/ITNG.2011.218

[42] Albash, T. & Lidar, D. A. Adiabatic quantum computation. Rev. Mod. Phys. 2018, 90, 015002.

[43] Kato, T. On the Adiabatic Theorem of Quantum Mechanics. Journal of the Physical Society of Japan. 1950, 5(6),435–439.

[44] Marzlin, K. P. & Sanders, B. C. Inconsistency in the application of the adiabatic theorem. Phys. Rev. Lett. 2004, 93, 160408.

[45] Amin, M. H. S. Consistency of the Adiabatic Theorem. Physical Review Letters. 2009, 102, 220401.

[46] Tong, D. M., Singh, K., Kwek, L. C., & Oh, C. H. Sufficiency Criterion for the Validity of the Adiabatic Approximation. Phys. Rev. Lett. 2007, 98, 150402.

[47] Grover, L. K. Quantum Mechanics helps in searching for a needle in a haystack. Phys. Rev. Lett. 1997, 78(2),325–328.

[48] Boyer, M., Brassard, G., Hoyer, P., & Tapp, A. Tight bounds on quantum searching. Fortschr. Phys. 1998, 46, 493–506.

[49] Tulsi, T. A., Grover, L. K., & Patel, A. A new algorithm for fixed point quantum search. Quantum information & computation. 2006, 6, 483–494. arXiv: quantph/0505007

[50] Dürr, C. & Hoyer, P. A Quantum Algorithm for Finding the Minimum. http://lanl.arxiv.org/abd/quant-ph/9607014

[51] Kenigsberg, D. Grover's Quantum Search Algorithm and Mixed States. M.Sc. Thesis. Israel Institute of Technology, October, 2001.

[52] Baritompa, W. P., Bulger, D. W., & Wood, G. R. Grover's Quantum Algorithm applied to Global Optimization. SIAM J. OPTIM. 2005, 15(4),1170–1184.

[53] Zabinsky, Z. B. & Smith, R. L. Pure adaptive search in global optimization. Math. Programming. 1992, 53, 323–338.

[54] Zabinsky, Z. B, Wood, G. R., Steel, M. A., & Baritompa, W. P. Pure adaptive search for finite global optimization. Math. Programming. 1995, 69, 443–448.

[55] Bulger, D. W., Alexander, D. L. J., Baritompa, W. P., Wood, G. R., & Zabinsky, Z. B. Expected hitting time for backtracking adaptive search. Optimization. 2004, 53, 189–202.

[56] Liu, Y., Koehler, G. J. Using modifications to Grover's Search algorithm for quantum global optimization. European Journal of Operational Research. 2010, 207(2),620–632.

[57] Michalewicz, Z. Genetic Algorithms + Data Structures = Evolution Programs. 3rd ed., Berlin, Springer-Verlag, 1996.

[58] Goldberg, D. Genetic Algorithms in Search, Optimization and Machine Learning. Addison-Wesley Publishing Company, Inc., NY, USA, 1989.

[59] Bellman, R. Adaptive Control Processes: A Guided Tour, Princeton University Press, 1961.

[60] Protopopescu, V., Barhen, J. Quantum Algorithms for Continuous Global Optimization. Proceedings of the 34th Workshop on Optimization and Control with Applications, Erice, Sicily, July 9–17, 2001.

Arit Kumar Bishwas, Ashish Mani, and Vasile Palade

4 From classical to quantum machine learning

Abstract: In recent years, Machine Learning (ML) has started to be ubiquitously applied to practically most of the human activity domains. Although traditional, or classical machine learning (CML) approaches are useful in solving many complex tasks, there are still many challenges that such approaches are facing. One issue is the limitation in the processing speed of current silicon technology based computers, which made researchers look to the underlying quantum theory principles and try to run complex quantum computing experiments. In future, in order to develop more complex artificial intelligence systems, we will have to process huge amounts of data at high speed, and the existing classical computing will probably not serve well this purpose, due to silicon technology limitations. A different technology that can handle huge volumes of data and at high speed is needed. In recent years, the progress in quantum computing research seems to provide hopeful answers to overcome the speed processing barrier. This is very important for the training of many computational intensive machine learning models. The latest advancements in quantum technology appear to be promising, which can boost the field of machine learning overall. In this chapter, we discuss the transition from classical machine learning to quantum machine learning (QML) and explore the recent progress in this domain. QML is not only associated with the development of high-performance machine learning algorithms that can run on a quantum computer with significant performance improvements but also has a very diverse meaning in other aspects. The chapter tried to touch those aspects in brief too, but the main focus is on the advancements in the field of developing machine learning algorithms that will run on a quantum computer.

Keywords: quantum algorithm, quantum random access memory (QRAM), quantum deep learning, quantum computation

4.1 Introduction

Machine learning is an important and popular field of study these days when Artificial Intelligence and intelligent algorithms have become embedded in our everyday life. Machine learning models are trained with large amounts of data using

Arit Kumar Bishwas, AIIT, Amity University, Noida, Uttar Pradesh, India
Ashish Mani, EEE, ASET, Amity University, Noida, Uttar Pradesh, India
Vasile Palade, Faculty of Engineering and Computing, Coventry University, Coventry, UK

https://doi.org/10.1515/9783110670707-004

some mathematical and statistical based algorithms. Machine learning algorithms are categorized into two broad classes – supervised and unsupervised learning algorithms. In supervised learning, data are provided in the form of labelled data. In the case of unsupervised learning, the datasets are not labelled. In recent years, machine learning has demonstrated great visibility and publicity in applications to many important domains of real life, such as medical image processing, online product recommendation systems, financial modelling and prediction, risk evaluation, human language processing, etc. The recent advancements in deep learning research opened even more possibilities for solving complex problems. In machine learning, data play a vital role. Some problems can be solved well with less data, but deep learning requires huge volumes of data. Apart from the issue of availability of big datasets, another crucial factor is the processing and learning speed of the system. Deep learning algorithms are computationally very intensive. Although by using GPUs and high-performance computing systems, this problem can be somehow alleviated, it still takes a long time to perform machine learning (and especially deep learning) on big data. The rule of thumb is that, in order to train with more data, we would require more processing power. To develop more complex future artificial intelligence systems that process a huge amount of data, the existing classical computing technologies will probably not be sufficient, due to the limitations related to the silicon-based technology. We need a different technology, which can handle the processing speed issue and which can process huge volumes of data. Classical computers are now part of our lives and help us in solving complex problems, but still, there are great challenges in solving some of the problems (for example optimization problems) whose computational complexities increases at a very high scale with a small increase in inputs. These kind of problems are very difficult or not possible to solve with the classical computers. Quantum computing is the answer to overcome these processing issues. Quantum computer's computational power scales significantly as the system size grows due to its unique properties which are not found in classical systems, and two of the most important properties are – superposition & entanglement. Section 3 discusses these properties in detail.

The term quantum machine learning (QML) mostly refers to classical machine learning algorithms [1, 2] that can be run on a quantum computer. This involves mostly researchers working on developing quantum versions of classical machine learning algorithms, concerned mainly with the speed-up factors for the training procedure. Apart from this, the term QML also has a general connotation. Developing a fully functional commercial quantum computer is a very active field, but it is still not clear if we can have one to work in the near future. In recent time, researchers focused on developing small quantum subroutines/devices which can combine with classical machine learning approaches to get the quantum advantages to some extent [3]. These quantum subroutines/devices can be designed to solve some of the complex tasks which are computationally difficult in the classical domain [4–6].

Some recent works have also exhibited that by using quantum algorithms we can analyze quantum states [7]. The term QML is also associated with analyzing the data generated from quantum experiments. Applying machine learning in analyzing quantum phase transitions is a very classic example of this [8–11].

Another active research field in QML is to understand the structural and operational similarities between certain physical systems and deep learning. Deep leaning uses some mathematical and numerical techniques that are inspired by quantum formulation, and in the same way, deep learning explores some structural and procedural equivalences with quantum systems [12–14].

Big companies like D-Waves [15], IBM [16], and Google [17] demonstrated some progress in building a quantum computer. Here, the D-Wave's model is based on the adiabatic quantum process [18] and the IBM [16] is working on mostly non-adiabatic models. IBM released the IBM-Q quantum computer to be accessed and used freely (with the restriction of 5 qubits system access at the time). Apart from the progress in quantum hardware, there is very active research going on in the area of developing quantum algorithms. Some of the most notable quantum algorithms are Grover's search [10], Shor's algorithm [19], HLL [20], etc., which demonstrated significant speed-up gains insolving computational problems. In recent years, the advancement in quantum technologies has involved some state-of-the-art machine learning approaches [21–25]. These quantum machine learning algorithms are exponentially faster than their classical counterparts.

In this chapter, our focus will be on a category of classical machine learning algorithms that can run on a quantum computer. At first, we discuss quantum computing basics in brief, including some well-known techniques that are useful in the advancement of quantum machine learning algorithms, followed by the recent advancements in the field of quantum machine learning algorithms.

4.2 Overview of quantum computing

4.2.1 Hilbert space & Dirac notation

A Hilbert space $\mathbb{C}^n$ is an abstract vector space which is very useful for representing a quantum system. The Dirac notation [26] is a very convenient way of defining vectors in the Hilbert space, named after the Nobel laureate Paul Dirac. In Dirac notation, a vector V is represented as:

$$V = \begin{bmatrix} v_0 \\ v_1 \\ \vdots \\ v_n \end{bmatrix} = |V\rangle \tag{4.1}$$

Here, $|V\rangle$ is known as "ket V". In the same context, the dual vector of $|V\rangle$ is known as "bra V" and represented as:

$$\langle V| = [\overline{v_0}\ \overline{v_1} \ldots \overline{v_n}] = \overline{V^T} \tag{4.2}$$

We define the inner product of vectors U and V in Dirac notation as:

$$\langle U|V\rangle = [\overline{u_0}\ \overline{u_1} \ldots \overline{u_n}] \begin{bmatrix} v_0 \\ v_1 \\ \vdots \\ v_n \end{bmatrix} = \overline{U^T} V \tag{4.3}$$

Also, the length of the vector in the Hilbert space, known as "Norm", is defined as:

$$\| |V\rangle \| = \sqrt{\langle V|V\rangle} \tag{4.4}$$

Similarly, the outer product can be defined as follows with the Dirac notation:

$$|V\rangle\langle U| = \begin{bmatrix} v_0 \\ v_1 \\ \vdots \\ v_n \end{bmatrix} [\overline{u_0}\ \overline{u_1} \ldots \overline{u_m}] = \begin{bmatrix} v_0\overline{u_0} & \cdots & v_0\overline{u_m} \\ \vdots & \ddots & \vdots \\ v_n\overline{u_0} & \cdots & v_n\overline{u_m} \end{bmatrix} \tag{4.5}$$

4.2.2 Quantum bits and gates

In classical computation, the fundamental unit for the building block of any computation is the bit. In the same context, the fundamental component of any quantum computation is known as a qubit. In classical computation, a base-2 number known as a bit can be either 0 or 1, but a qubit can also be in both states, 0 and 1, at the same time. Mathematically we represent a qubit as a two-dimensional state space in $\mathbb{C}^2$ with orthogonal basis vectors $|0\rangle$ & $|1\rangle$ as,

$$|\psi\rangle = a_0|0\rangle + a_1|1\rangle; |a_0|^2 + |a_1|^2 = 1 \tag{4.6}$$

The states $|0\rangle$ & $|1\rangle$ are in superposition (this will be covered in detail in the next section) and the complex scalar amplitudes a_0 & a_1 are the probabilities of obtaining the values $|0\rangle$ & $|1\rangle$ upon measurement, respectively.

A single qubit is of no significant use, but a bunch of qubits can involve much more practical usability. Like a classical computer, the quantum register is the collection of qubits. In the quantum register, a qubit in a register can be in a superposition of $|0\rangle$ & $|1\rangle$, and upon measurement collapse to a bit. So, if there is a n qubits register, then there will be 2^n possible states in superposition. For example, suppose we have a 3 qubits register, then there will be $2^3 = 8$ states in superposition as shown:

$$|\psi_3\rangle = \sum_{i=0}^{2^3-1} a_i|i\rangle = a_0|000\rangle + a_1|001\rangle + a_2|010\rangle + a_3|011\rangle + a_4|100\rangle + a_5|101\rangle + a_6|110\rangle + a_7|111\rangle \tag{4.7}$$

Like classical computing, we have quantum gates too, which are used to perform some kind of operations on qubits and registers. The followings are some of the most important quantum gates:

a. Hadamard gate (H) – This is a very important quantum gate. The Hadamard gate transforms a qubit into a superposition of qubits as shown:

$$\mathrm{H}|0\rangle = \frac{|0\rangle + |1\rangle}{\sqrt{2}} \tag{4.8}$$

$$\mathrm{H}|1\rangle = \frac{|0\rangle - |1\rangle}{\sqrt{2}} \tag{4.9}$$

b. Pauli-X quantum gate (X) – Pauli–X is quantum equivalent to classical NOT gate. It swaps the amplitudes of the $|0\rangle$ & $|1\rangle$.
c. Pauli-Y quantum gate (Y) – It swaps the amplitudes of $|0\rangle$ & $|1\rangle$, multiplies each amplitude by i, and then negates the amplitude of $|1\rangle$.
d. Pauli-Z quantum gate (Z) – When applying a Pauli-Z quantum gate to a qubit, it negates the amplitude of the qubit if it is $|1\rangle$, otherwise, it performs no action if the qubit is $|0\rangle$.

4.2.3 Superposition

The Quantum superposition [27] is a phenomenon in the quantum world in which quantum particles appear to exist in multiple states simultaneously. It is the fundamental principle of quantum mechanics. When the particle is measured in superposition, the superposition collapses and results in correspond to only one of the possible states.

Thus, at any instance, a qubit can be in a superposition of both states $|0\rangle$ and $|1\rangle$ simultaneously, such as:

$$|W\rangle = a|0\rangle + b|1\rangle = \begin{pmatrix} a \\ b \end{pmatrix} \quad (4.10)$$

where the coefficients a and b are complex numbers and the amplitudes of the components $|0\rangle$ and $|1\rangle$ respectively, possessing the following property:

$$|a|^2 + |b|^2 = 1 \quad (4.11)$$

4.2.4 Entanglement

Quantum entanglement [28] is a very important property of quantum mechanics, in which qubits interact with each other instantaneously irrespective of the distances between them through an unknown way of communication that is not limited to the speed of light. Correlated particles remain entangled as long as they are isolated irrespective of the distance between them.

If we cannot factor a multi-qubit state into the direct product of a definite state for each qubit individually, then the states are entangled. Hence, a pair of qubits A and B are entangled if and only if we cannot write the joint state $|\psi_{AB}\rangle$ as the product of a state for qubit A and a state for qubit B, i.e., if and only if $|\psi_{AB}\rangle = |\psi_A\rangle \otimes |\psi_B\rangle$ for any choice of states $|\psi_A\rangle$ and $|\psi_B\rangle$.

We are considering two non-interacting systems A and B, with respect to finite dimensional Hilbert spaces H_A and H_B.

$H_A \otimes H_B$ is the tensor product of the composite system in the Hilbert space. The state of the composite system is $|\psi_A\rangle \otimes |\psi_B\rangle$, where the first system is in state $|\psi_A\rangle$ and the second in the state $|\psi_B\rangle$.

The above represented states of the composite system are separable states. But, all states are not, in general, the separable states.

By fixing a basis $|i_A\rangle$ for H_A and a basis $|j_B\rangle$ for H_B, the most general state in $H_A \otimes H_B$ is of the form:

$$|\psi_{AB}\rangle = \sum_{i,j} c_{ij}|i_A\rangle \otimes |j_B\rangle \quad (4.12)$$

This state is separable if

$$c_{ij} = c_i^A c_j^B \quad (4.13)$$

yielding:

$$\left|\psi_A\right\rangle = \sum_{i,j} c_i^A \left|i_A\right\rangle \text{ and } \left|\psi_B\right\rangle = \sum_j c_i^B \left|j_B\right\rangle \tag{4.14}$$

It is inseparable if

$$c_{ij} \neq c_i^A c_j^B \tag{4.15}$$

If a state is not separable, it is known as entangled state.

For example, the following composite state is an entangled state:

$$\frac{1}{\sqrt{2}}(|0_A\rangle \otimes |1_B\rangle - |1_A\rangle \otimes |0_B\rangle), \tag{4.16}$$

where $|0_A\rangle$ & $|1_A\rangle$ are two basis vectors of H_A and $|0_B$ & $|1_B$ are two basis vectors of H_B.

4.2.5 Quantum adiabatic model

If a quantum system is set in the ground state of a Hamiltonian H, the system remains in this state. According to the adiabatic theorem [18], the system will still stay close to the ground state as long as the variation is very slow enough. Suppose that $|E_0; t\rangle$ & $|E_1; t\rangle$ are the ground and first excited states of the Hamiltonian $H(t)$, where $E_0(t)$ and $E_1(t)$ are the energies corresponding to the ground and first excited states, respectively. We then define the minimum gap between these two eigenvalues as:

$$g_{min} = \min_{0 \leq t \leq T} [E_1(t) - E_0(t)] \tag{4.17}$$

So, when we prepared the system at time $t = 0$ in its ground state $|E_0; 0$ and allow it to evolve under the Hamiltonian $H(t)$, we find:

$$|\langle E_0;T|\psi(T)\rangle|^2 \geq 1 - \varepsilon^2 \tag{4.18}$$

where $1 \gg \varepsilon \geq \frac{\max_{0 \leq t \leq T} \left|\left\langle \frac{dH}{dt} \right\rangle_{1,0}\right|}{g_{min}^2}$, and $\left\langle \frac{dH}{dt} \right\rangle_{1,0} = \left\langle E_1; t \left| \frac{dH}{dt} \right| E_0; t \right\rangle$

4.2.6 SWAP-test

A SWAP-test estimates $|\langle\psi_1|\psi_2\rangle|^2$, when given two input quantum states $|\psi_1\rangle$ and $|\psi_2\rangle$, with precision $O\left(\frac{1}{\sqrt{M}}\right)$, where M is the number of times the measurement has been done.

4.2.7 Quantum algorithms

Quantum algorithms are the special type of algorithms that are written for running on a quantum computer. These algorithms are based on the principles of quantum mechanics. This is one very active research area, and recent research works show that many of the quantum versions of classical algorithms are way more efficient in terms of computational complexity, with some of them being exponentially faster than the classical counterparts. Some of the most known quantum algorithms are Grover's search algorithm [10], Shore's algorithm [19], quantum support vector machine [21, 23, 24], and the HHL algorithm [20].

4.3 Essential foundations for building a quantum machine learning framework

In this section, we discuss the essential foundations to develop quantum machine learning algorithms. These foundational concepts are very important. No single concept can be used to develop a quantum machine algorithm, and different concepts are used to cultivate different algorithms. In this section, we explain some of the very key foundational concepts which provide support in building quantum machine learning algorithms. We have discussed how different approaches used these concepts to articulate a variety of quantum machine algorithms in Section 4.4 & 4.5.

4.3.1 Grover's search algorithm

Grover's search algorithm [10] is a quantum search algorithm which performs the search to find a unique item from an unordered set of N items with $O(\sqrt{N})$ run time complexity. The best counter classical search algorithm performs with $O(N)$ runtime complexity. The algorithm is as follows [29]:

Algorithm 1: Grover's Search Algorithm
1. Initialize quantum state $|0\rangle^{\otimes n}$
2. Apply the Hadamard transformation to all the qubits

$$|\psi\rangle = H^{\otimes n}|0\rangle^{\otimes n} = \frac{1}{\sqrt{2^n}}\sum_{y=0}^{2^n-1}|y\rangle$$

3. Apply the Grover iteration function R $\approx \frac{\pi}{4}\sqrt{2^n}$ times

$$[(2|\psi\rangle\langle\psi| - I)O]^R|\psi\rangle \approx |y_0$$

4. Perform the measurement for output, y_0

4.3.2 Quantum random access memory

Quantum random access memory (QRAM) [30] is the random access memory in a quantum paradigm. It encompasses the output and the address registers which are composed of qubits. The QRAM permits to access the data quantum mechanically and access the memory in coherent quantum superposition. The address register embraces a superposition of addresses. The data register *DR* stores the output of the QRAM as a superposition of data, which is associated with the address register R_{ADR},

$$\sum_j \psi_j|j\rangle_{R_{ADR}} \rightarrow \sum_j \psi_j|j\rangle_{ADR}|D_j\rangle_{DR} \tag{4.19}$$

where R_{ADR} holds a superposition of addresses $\sum_j \psi_j|j\rangle_{R_{ADR}}$, and D_j is the jth memory cell content. Reconstructing any quantum state from QRAM takes $O(log\ N)$ steps, where N is the complex vector dimension.

4.3.3 Quantum dot product

We discuss a quantum-based process for dot product evaluation [31] of two training inputs (normalized), $\|\vec{X}_i\rangle$ & $\|\vec{X}_j\rangle$ (in the quantum arrangement), in a linear kernel. For calculating a dot product of $\|\vec{X}_i\rangle$ & $\|\vec{X}_j\rangle$, at first, we generate two quantum states $\|\psi\rangle$ and $\|\phi\rangle$ using an ancilla variable. Next, we estimate the squared norms sum of the two training inputs, say parameter $Z = |\vec{X}_i|^2 + |\vec{X}_j|^2$. At last, we execute a swap test to implement a projective measurement on the ancilla alone.

Initially, by querying the QRAM, we construct a quantum state $|\psi\rangle$:

$$|\psi\rangle = \frac{1}{\sqrt{2}}\left(|0\rangle|\vec{X}_i\rangle + |1\rangle|\vec{X}_j\rangle\right) \tag{4.20}$$

Let us assume another quantum state:

$$|\xi\rangle = \left(\frac{1}{\sqrt{2}}(|0\rangle - |1\rangle) \otimes |0\rangle\right) \tag{4.21}$$

We put on a unitary transformation [32] e^{-iHt} to the state $|\xi\rangle$, where $H = ((\|\vec{X}_i\| |0\rangle\langle 0| + \|\vec{X}_j\| |1\rangle\langle 1|) \otimes \sigma_x)$ is a Hamiltonian. This outcomes the following state,

$$\left[\frac{1}{\sqrt{2}}(cos(\|\vec{X}_i\|t)|0\rangle - cos(\|\vec{X}_j\|t)|1\rangle) \otimes |0\rangle\right] - \left[\frac{i}{\sqrt{2}}(sin(\|\vec{X}_i\|t)|0\rangle - sin(\|\vec{X}_j\|)|1\rangle) \otimes |1\rangle\right] \tag{4.22}$$

With an appropriate choice of t, we now measure the ancilla bit, where $\vec{X}_i t, \vec{X}_j t \ll 1$, which outcomes in the state:

$$|\phi\rangle = \frac{1}{\sqrt{(\|\vec{X}_i\|^2 + \|\vec{X}_j\|^2)}}(\|\vec{X}_i\| |0\rangle - \|\vec{X}_j\| |1\rangle), \tag{4.23}$$

with probability $\frac{1}{2}[(\|\vec{X}_i\|^2 + \|\vec{X}_j\|^2)t]^2$.

This sanctions the estimation of the squared norms sum of $|\vec{X}_i\rangle$ & $|\vec{X}_j\rangle$. By applying quantum counting, we can estimate $(\|\vec{X}_i\|^2 + \|\vec{X}_j\|^2)$ and generate the quantum state $|\phi\rangle$ with accuracy ϵ. The complexity will be $O(\epsilon^{-1})$. Using an ancilla alone, we then execute a swap test with states $|\psi\rangle$ & $|\phi\rangle$. If $|\psi\rangle$ & $|\phi\rangle$ are equal then the measurement gives us a zero. Thus, the overall complexity in evaluating a single dot product of the training instances, considering the QRAM access [30], estimating $(\|\vec{X}_i\|^2 + \|\vec{X}_j\|^2)$, and fabricating the quantum state $|\phi\rangle$ is $O(\epsilon^{-1} \log N)$.

4.3.4 The HHL algorithm

Harrow, Hassidim, and Lloyd [20] developed the HHL algorithm which quantum mechanically inverts a system of linear equations. This algorithm is one of the most significant and fundamental quantum algorithms, and the base for constructing many QML algorithms. The algorithm pursues to solve the system of equations $A\vec{x} = \vec{b}$ in a quantum computer. Quantum mechanically it is represented as $A|x\rangle = |b\rangle$, where A is a Hermitian matrix. $\vec{b}$ is expressed as a quantum state $|b\rangle = \sum_n b_n|E_n\rangle$ over $log_2 N$ qubits where E_n is an eigenvector of A with eigenvalue $\lambda_n \geq \Lambda$, and $\vec{x}$ is stated as $|x\rangle$. The algorithm aims to solve $|x\rangle = A^{-1}|b\rangle$, where A^{-1} is the inverse of the A. We then compute λ_n by applying the phase estimation. By uncomputing the phase-estimation we get $\sum_n b_n|E_n\rangle\left(\frac{\Lambda}{\lambda_n}|1\rangle + \sqrt{1 - \frac{\Lambda^2}{\lambda_n^2}}|0\rangle\right)$, where $\frac{\Lambda}{\lambda_n}$ is the angle of arcsin through which an ancillary qubit is rotated. To succeed, we apply $O(||A||/\Lambda)$ number of times the state preparation circuit after applying amplitude amplification.

As compared to the classical counterpart, the algorithm claims exponential speed-up gain under certain problem statements. The algorithm takes $O((logN)^2)$

quantum states to output, whereas it takes $O(NlogN)$ steps to output in a classical computer.

4.3.5 Quantum principal component analysis

Principal component analysis (PCA) [33] is a very important technique for dimensionality reduction. In a more formal way, we can define that PCA operates by diagonalizing the covariance matrix $C = \sum_j \vec{v}\vec{v}_j^T$ of the data, where T is the transpose operation. The covariance matrix encapsulates the correlations between data's different components. The covariance matrix in the form of eigenvectors and eigenvalues can be represented as $C = \sum_k e_k \vec{c}_k \vec{c}_k^{\dagger}$, where e_k are the eigenvalues correspond to the eigenvectors $\vec{c}_k$ of C. The few eigenvalues which are large as compared to the rests, are known as the principal components. Each of the principal components are then considered as the new feature vectors. The classical algorithm for PCA works in $O(d^2)$ runtime complexity, where d is the dimension of the Hilbert space. In quantum PCA [33], a randomly chosen data vector is mapped into a quantum state using QRAM [30]. The resulting state has a density matrix $\rho = \frac{1}{N}\sum_j |v_j\rangle\langle v_j|$, where N is the cardinality of the data vector set. Using density matrix exponentiation [34] with the quantum phase estimation algorithm [35], and repeating the data sampling, allow us to decompose the data vectors into principal components. The quantum PCA algorithm works in $O((logd)^2)$ runtime complexity.

4.4 Quantum machine learning approaches

4.4.1 HHL or quantum amplitude based approaches

The objective of amplitude encoding based machine learning algorithms [9, 11, 20, 33, 36–39, 40] is to articulate quantum machine learning algorithms whose resources grow in polynomial runtime with the n qubits, and expands to a logarithmic growth with amplitudes and thus the input dimension. Many QML algorithms based on this approach uses the speedup advantages by solving a system of linear equations quantum mechanically [20]. Quantum matrix inversion is one of the main core processes in solving the system of linear equations exponentially faster than the classical counterpart. Some of the notable quantum machine learning algorithms based on amplitude encoding approach are the quantum support vector machine [21–23] and least-square linear regression [37, 38].

4.4.2 Grover's search-based approaches

Machine learning algorithms based on Grover's search algorithms, which use the technique of amplitude amplification to unfold an unstructured search problem with a quadratic runtime speedup, apply Grover's search subroutine to solve a larger problem. Quantum K-Means [25], K-Median [41], and k-nearest neighbors [42] are some of the machine learning algorithms which use Grover's search algorithms as a subroutine to solve the machine learning task. Another recent use of the Grove's search subroutine is in training the perceptron [43].

4.4.3 Quantum annealing based approaches

Very recently, some experiments have been done based on quantum annealing [44]. In this work, the researchers used a quantum annealer to classify and rank the transcription factors binding to study how the proteins (in a gene) identify and specifically bind to their DNA targets. The researchers used the D-Wave system to implement the quantum annealing algorithm. Here, the machine learning techniques have been the effective tools to disclose interaction mechanism at the genetic level.

4.5 Recent progress in quantum machine learning algorithms

The research in developing quantum machine learning algorithms has exponentially grown in recent years, and day by day new interesting quantum machine learning algorithms are being published. Here, we are covering the most ground-breaking and popular algorithms which demonstrated significant state-of-the-art performances as compared to their classical counterparts.

4.5.1 Quantum binary least square support vector machine

The quantum binary least square support vector machine is a break-through in the field of QML, where the authors, Rebentrost et al. [21], demonstrated an exponential speedup gain in support vector machine formulation.

In the least square SVM, the optimization problem is transformed into a system of linear equations which encompasses the kernel matrix for solving the optimization problem. We solve the system of linear equations in a much faster way by solving the least squares formulation and speeding up the kernel matrix calculations.

The following areas where we take the quantum advantages to improve the performance of the SVM are:

a. Data preparation with quantum random access memory (QRAM) – One of the challenging tasks in quantum machine learning is to prepare the data set for feeding into the quantum systems for processing. QRAM helps in preparing the classical data set into the quantum form. QRAM takes $O(log_2\, d)$ steps to query the information from memory for reconstructing a state, where d is the dimension of the feature vector.
b. Kernel matrix calculation – The dot product evaluation plays the main role in the kernel matrix calculation. So, if we get the speed up advantages in the dot product evaluation in the quantum formulation, it will imply overall speedup gain in the kernel matrix calculation. In [35], authors have discussed the faster way to do dot product calculation with the help of quantum properties. Also, the exponentiation of the inverse of the normalized kernel matrix K^{-1} is performed in a quantum paradigm. As we have discussed earlier, QRAM takes only $O(log_2 d)$ steps to query for reconstructing a state, thus a single dot product with QRAM takes $O\,(\epsilon^{-1}\, log_2\, d)$ steps, where ϵ is the accuracy.
c. Least square formulation – The speedup gain is possible during the training phase because of the quantum mechanical implementation of the exponentially faster eigenvector formulation in the non-sparse density matrices [35], the sparse matrix simulation [45], and the matrix inversion algorithm [35].

The least square SVM is represented as the following formulation:

$$\hat{F}(|b,\ \vec{\alpha}\rangle) = |\vec{y}\rangle,\ \hat{F} = \begin{bmatrix} 0 & 1^T \\ 1 & K + \gamma^{-1} I \end{bmatrix}, \tag{4.24}$$

where $\hat{F} = \frac{F}{trF}$ is the normalized F, and trF is a trace of F. The target is to solve the optimization problem, $\hat{F}(|b,\ \vec{\alpha}\rangle) = |\vec{y}\rangle$, for the parameters b and $\vec{\alpha}$. Here, K is the kernel matrix.

4.5.2 Quantum multiclass support vector machine

The SVM is a binary class classification algorithm. To handle multiclass classification, we use *one-against-all* and *all-pair* algorithms. The quantum multiclass support vector machine (SVM) is the generalized version of the quantum binary SVM. In [23], the authors discussed two techniques: the quantum version of the *one-against-all* as well as the quantum version of *all-pair* algorithms. In *one-against-all*, we first construct and train k binary classifiers and then each of these classifiers classify a given unknown input. The class for which the corresponding classifier's probability confidence score is the highest, is selected as the predicted class. In *all-pair*, we construct

$k(k-1)/2$ classifiers. Each classifier is trained with a dataset which contains one primary class data and the rest of the data for another class, i.e., we make pairwise datasets and train the classifiers, resulting in $k(k-1)/2$ total classifiers. When an unknown input is to get predicted, these all $k(k-1)/2$ classifiers predict the class for the given input and by applying voting mechanism we select the class to be predicted. The following algorithms show the quantum version of these two approaches, for detail explanations please refer to [22, 23].

Algorithm 2: Quantum One-Against-All Algorithm

1. *initialize the index* $I_{index} = $ *any randon element*, $1 \leq I_{index} \leq k$
2. *initialize* $|V_{qn}\rangle$ *as the vector of all classified class probabilities in QRAM*
3. *while* $(P_m < P_{threshold})$
4. *initialize the memory as* $|\Psi\rangle = \frac{1}{\sqrt{k}} \sum_r |r\rangle |I_{index}\rangle$
5. *GROVER-QUANTUM-SEARCH* $(|\Psi\rangle, |V_{qn}\rangle, I_{index})$
6. *if* $(|V_{qn}\rangle[I_{index\_new}] > |V_{qn}\rangle[I_{index}])$
7. $I_{index} = I_{index\_new}$
8. *return* I_{index}

Algorithm 3: Quantum All-Pair Algorithm

1. *initialize class* = *any random element*, $1 \leq class\_index \leq k$
2. *initialize* $|V_q\rangle$ *as the vector of all classified classes*
3. *initialize frequency estimate* s_{class_index} *with any very small value*
4. *INITIAL-FREQUENCY-COUNT* $(class\_index,\ s_{class_{index}})$
5. *while* $(total\ running\ time < O\ (\log k))$
6. *initialize the memory* $|C_i\rangle = \sum_j \beta_{i,j} |j\rangle$
7. *initialize the memory as*

$$|\psi_c\rangle = \sum_{i=0}^{k-1} |i\rangle |C_i\rangle |class\_index\rangle |s_{class\_index}\rangle$$

8. *GROVER-QUANTUM-SEARCH* $(|\psi_c\rangle, |V_q\rangle,\ class\_index,\ s_{class\_index})$
9. *MEASURE-REGISTER* $(|\psi_c\rangle)$
10. *if* $\left(s_{class\_index\_new} > s_{class\_index} + \frac{\varepsilon}{2k}\right)$
11. $class\_index = class\_index_{new}$
12. $s_{class\_index} = s_{class\_index\_new}$
13. *return* $|V_q\rangle[class\_index]$

4.5.3 Quantum K-means clustering

Clustering is a very important problem in machine learning and more generally falls under unsupervised learning. Clustering algorithm groups the unlabeled dataset into multiple categories based on their similar features. The K-Means clustering algorithm is a very popular and simple clustering algorithm. K-Means groups the unlabelled dataset into k categories, however, we have to specify K in advance. In [25], authors show the formulation of a K-Means clustering algorithm in the quantum paradigm. In the quantum K-Means, the closest centroids are calculated in the same way as in the classical version, but by using the Grover's search [10]. The quantum version of k-Means has been claimed to be exponentially faster in runtime complexity as compared to the classical counterpart. The quantum K-Means is based on an adiabatic quantum process.

Suppose the output of the quantum K-Means clustering algorithm is in the following quantum state:

$$|\xi\rangle = M^{-1/2} \sum_{l} |c_l\rangle |l\rangle = M^{-1/2} \sum_{c, l\epsilon c} |c\rangle |l\rangle \tag{4.25}$$

We construct the above output quantum state with $O\left(\epsilon^{-1} K \, log(KMN)\right)$ runtime complexity, where ϵ is the accuracy, and with $O\left(\epsilon^{-1} log\left(KMN\right)\right)$ runtime complexity when the clusters are well separated. The adiabatic gap in this case is $O(1)$.

Let us first select K objects with labels i_c as initial seeds. We begin with the state:

$$\left\{(MK)^{-1/2}\right\} \sum_{c'l} |c'\rangle |l\rangle \left\{\left(K^{-1/2}\right) \sum_{c} |c\rangle |i_c\rangle\right\}^{\otimes D} \tag{4.26}$$

Here, D copies of the state permit us to evaluate the distances $|\vec{x}_l - \vec{x}_{i_{c'}}|^2$ in the $c'l$ superposition component. By applying the adiabatic algorithm with the initial Hamiltonian

$$H_0 = 1 - |\varphi\rangle\langle\varphi|, |\varphi\rangle = K^{-0.5} \sum_{c'} |c'\rangle \tag{4.27}$$

The initial clustering is:

$$|\psi_1\rangle = M^{-1/2} \sum_{c, l\epsilon c} |c\rangle |l\rangle \tag{4.28}$$

We evolve the initial Hamiltonian H_0 slow enough to the Hamiltonian:

$$H_2 = \sum_{c'l} \left\{ \left|\vec{x}_l - \vec{x}_{i_{c'}}\right|^2 \right\} |c'\rangle\langle c'| \otimes |l\rangle\langle l| \tag{4.29}$$

We can construct the separate cluster $|\varphi_1^c = M^{-0.5}\sum_{l\epsilon c}|l\rangle$, assuming we have D copies of $|\psi_1\rangle$. We can now estimate the number of quantum states M_c in the cluster c. With D copies of $|\psi_1\rangle$ together with the clusters $|\varphi_1^c\rangle$, we evaluate the average distance between x_l and the individual cluster's mean

$$\left|\vec{x}_l - (M_c)^{-1}\sum_{K\epsilon c'}\vec{x}_K\right|^2 = |\vec{x}_l - \vec{x}_c|^2, \tag{4.30}$$

by applying a phase $e^{-i|\vec{x}_l - \vec{x}_{c'}|^2\delta t}$ to each component $|c'|l\rangle$ of the superposition using the following Hamiltonian,

$$H_F = \sum_{c'l}\left\{|\vec{x}_l - \vec{x}_{c'}|^2\right\}|c'\rangle\langle c'| \otimes |l\rangle\langle l| \otimes I^D \tag{4.31}$$

With the initial Hamiltonian $H_0 = 1 - |\varphi\rangle\langle\varphi|$ at the ground state, we initiate the adiabatic evolution on the state $\{(MK)^{-0.5}\}\sum_{c',\,l}|c'|l\rangle|\Omega_1\rangle^{\otimes D}$, where $|\varphi\rangle = K^{-0.5}\sum_{c'}|c'\rangle$ is the superposition of the cluster centroids. We get the final state:

$$\left\{(M)^{-0.5}\sum_{c'l\epsilon c'}|c'\rangle|l\rangle\right\}|\Omega_1\rangle^{\otimes D} = |\Omega_2\rangle|\Omega_1\rangle^{\otimes D} \tag{4.32}$$

By repeating the final state D times, we construct D copies of $|\Omega_2$ and keep on iterating the cluster assignment procedure until it gets converged to $|\xi\rangle$.

4.5.4 Quantum deep learning

Deep learning (DL) revolutionized the world of machine learning with its performance for a large domain of applications. DL is an efficient and effective tool for machine learning modeling. With the development of GPUs, a very wide area opened for developing highly complex machine learning models using deep learning, which was not possible earlier without large availability of high-performance computing. Deep learning requires a large amount of training data. Processing such a huge volume of datasets is a computationally complex task, this inspired the exploration of the quantum advantages for DL algorithms. Recent progress in developing programmable photonic circuits and quantum annealers [46–48] open up the possibilities of developing a quantum framework for deep learning algorithms.

The latest research on Boltzmann machine formulation in a quantum paradigm [49] demonstrated very strong footsteps towards quantum deep learning. The current results are focusing on developing quantum deep learning algorithms which do not require a fully functional quantum computer. These quantum DL algorithms can run with quantum annealers, which are special-purpose quantum information processors and they are very easy to construct as compared to larger quantum computers.

4.6 Recent progress in the implementation of quantum machine learning

We discuss here some of the significant developments on implementing quantum machine learning algorithms.

In 2003 [50], a group of researches discussed that image data can be very efficiently encoded in quantum states. This may reduce the resources to some order of magnitude to perform some specific and challenging image processing tasks. During the year of 2009 [51], D-Wave demonstrated some remarkable results with their adiabatic D-Wave quantum system. Using regularized boosting with a non-convex objective function, researchers conducted experiments in the D-Wave system to identify cars in images. In the same year (2009), another research group showed mapping an input data and memorized the data with a Hamiltonian using a quantum Hopfield network with adiabatic quantum computation [52].

Since 2013 [53, 54], NASA, Google research, and the University Space Research Association are working on exploring the D-Wave quantum computer for solving complex optimization problems, which may have an impact in solving some of the machine learning algorithms with the help of advantages offered by quantum properties. In 2015, researchers demonstrated the classification between the digit '6' and '9' by implementing a quantum support vector machine algorithm with an NMR (Nuclear magnetic resonance) technology on a liquid-state quantum computer [55]. Another important result came in the same year 2015 when a research team implemented an all-optical linear classifier using non-linear photonics [56].

Recently, IBM launched an online framework for access to their quantum computer (named IBM Q), and the system supports around 32 + qubits to work with. They allow free access up to 5 qubits to work on the real quantum computer. We can run small specific quantum routines (e.g., Grover's search, Hadamard gate, quantum SWAP test, etc.) into this system, which can help in solving complex machine learning use cases by using the outputs of these specific small quantum subroutines [16]. A fully functional quantum neural network is possible by implementing nonlinear interactions in the quantum dynamics, recently claimed by a research team in 2016 [57].

Very recently, in 2017 [58], the researchers demonstrated that their model is able to generate handwritten digits. The model is a probabilistic generative model based on arbitrary pairwise connectivity. Also in 2017, researchers encoded image information with the quantum image in a quantum system and discussed some interesting challenges on this topic [59]. In 2017, a very interesting research work by Yao et.al. discussed the implementation of quantum image processing and its applications for detecting edges [60].

In July 2018, Google announced the first public alpha of Cirq, an open source framework at the First International Workshop on Quantum Software and Quantum

Machine Learning (QSML). This will allow developers to build quantum algorithms, which may include some simple quantum machine learning algorithms with small scale data sets, without requiring a quantum physics background [61]. In 2020, Google announced TensorFlow Quantum which is a quantum machine learning library for rapid prototyping of hybrid quantum-classical ML models [62].

4.7 Conclusion

Quantum machine learning is a fast emerging field of research that opens new horizons for artificial intelligence based research and applications. We are seeing active research developments in this area, and with the advancement of quantum computers and quantum devices, it may solve some of the most challenging problems that are difficult to solve with classical machine learning like clustering and classification training with big data in reasonable time period. Recent advancements in deep learning helped us train complex machine learning models with the help of GPUs and high-performance computing, which demonstrated remarkable speed ups and improved accuracies. More complex machine learning problems require high volumes of data and tremendous computational power, and, due to the limitations of silicon-based chip technology, we have significant constraints on the upper bound of the computational power. Quantum based systems promise significant improvements in speed up factors as compared to the classical systems. Some of the quantum machine learning algorithms, such as the quantum binary/multiclass SVM, quantum K-Means clustering, and the quantum Boltzmann machine, have revealed substantial performance improvements with big data. However, we have important challenges that need to be addressed in order to make effective and efficient quantum machine learning algorithms, such as: how to transform classical big data into quantum form in a very efficient way; the hardware where such quantum algorithms will be run is still at initial stages of development, so proper testing and validation of these theoretical quantum machine learning models is still an issue. Also, the QML opens new doors for exploring other domains like understanding the processing of neural network based algorithms like deep learning, understanding the quantum experiments data analysis with the classical machine learning. Small quantum devices are easier to develop as compared to fully functional quantum computers. It is believed these small specific quantum devices can speed up the merging of quantum terminologies and methods with the machine learning ones and can solve many domain-specific complex problems in a hybrid framework, where such small quantum devices are used to solve a small complex task in a big complex problem domain.

References

[1] Alpaydin, E. Introduction to Machine Learning. Adaptive Computation and Machine Learning. MIT Press, Cambridge, MA, 2004.

[2] Wikipedia, cite_note 5-2.

[3] Schuld, M., Sinayskiy, I., & Petruccione, F. An in-troduction to quantum machine learning. Contemporary Physics. 2014, 56(2),172–185.

[4] Benedetti, M., Realpe-Gómez, J., Biswas, R., & Perdomo-Ortiz, A. Quantum-Assisted Learning of Hard-ware-Embedded Probabilistic Graphical Models. Physical Review X. ISSN 2160-3308. 2017, 7(4),041052. arXiv:1609.02542. doi:10.1103/PhysRevX.7.041052

[5] Farhi, E. & Neven, H. Classification with Quantum Neural Networks on Near Term Processors. 2018-02-16. arXiv:1802.06002.

[6] Schuld, M., Bocharov, A., Svore, K., & Wiebe, N. Circuit-centric quantum classifiers. 2018-04-02. arXiv:1804.00633.

[7] Yu, S., Albarran-Arriagada, F., Retamal, J. C., Wang, Y. -T., Liu, W., Ke, Z.-J., Meng, Y., Li, Z.-P., and Tang, J.-S. Reconstruction of a Photonic Qubit State with Quantum Rein-forcement Learning. 2018-08-28.

[8] Broecker, P., Assaad, F. F., & Trebst, S. Quantum phase recognition via unsupervised machine learning. 2017-07-03. arXiv:1707.00663

[9] Huembeli, P., Dauphin, A., & Wittek, P. Identifying Quantum Phase Transitions with Adversarial Neural Networks. Physical Review B. 2018, 97(13),134109. arXiv:1710.08382. doi:10.1103/PhysRevB.97.134109. ISSN 2469-9950

[10] Grover, L. K. A fast quantum mechanical algorithm for database search. Proceedings, 28th Annual ACM Symposium on the Theory of Computing. May 1996, 212.

[11] Melnikov, A. A., Nautrup, H. P., Krenn, M., Dunjko, V., Tiersch, M., Zeilinger, A., & Briegel, H. J. Active learning machine learns to create new quantum experiments. Proceedings of the National Academy of Sciences. ISSN 0027-8424. PMC 5819408. PMID 29348200. 2018. 115(6), 1221–1226. arXiv:1706.00868. doi:10.1073/pnas.1714936115

[12] Huggins, W., Patel, P., Whaley, K. B., & Stoudenmire, E. M. Towards Quantum Machine Learning with Tensor Networks. 2018. arXiv:1803.11537

[13] Carleo, G., Nomura, Y., & Imada, M. Constructing exact representations of quantum many-body systems with deep neural networks. 2018-02-26. arXiv:1802.09558

[14] Bény, C. Deep learning and the renormalization group. 2013-01-14. arXiv:1301.3124

[15] https://www.dwavesys.com/quantum-computing

[16] https://quantumexperience.ng.bluemix.net/qx/experience

[17] https://ai.google/research/teams/applied-science/quantum-ai/

[18] Adiabatic Quantum Computation. Adiabatic quantum computation, Wikipedia, the free encyclopedia, 2016-06-20.

[19] Shor, P. W. AT&T Research. arxiv.org/abs/quant-ph/9508027

[20] Harrow, A. W., Hassidim, A., & Lloyd, S. Quantum Algorithm for Linear Systems of Equations. Phys. Rev. Lett. 2009, 103, 150502.

[21] Rebentrost, P., Mohseni, M., Lloyd, S. Quantum support vector machine for big data classification. 2014. arXiv:1307.0471v3

[22] Bishwas, A. K. Big Data Classification with Quantum Multiclass SVM and Quantum One-Against-All Approach. Conference Proceedings to IEEE 2nd International Conference on Contemporary Computing and Informatics (IC3I). 2016. DOI: 10.1109/IC3I.2016.7918805

[23] Bishwas, A. K. An All-Pair Approach for Big Data Multiclass Classification with Quantum SVM. Springer Nature's Quantum Information Processing Journal. September 2018. doi.org/10.1007/s11128-018-2046-z

[24] Bishwas, A. K. Quantum Supervised Clustering Algorithm for Big Data. IEEE 2018 3rd International Conference for Convergence in Technology (I2CT). April 2018. DOI: 10.1109/I2CT.2018.8529799
[25] Lloyd, S., Mohseni, M., Rebentrost, P. Quantum algorithms for supervised and unsupervised machine learning. 2013. arxiv.org/abs/1307.0411
[26] https://en.wikipedia.org/wiki/Bra%E2%80%93ket_notation
[27] https://en.wikipedia.org/wiki/Quantum_superposition
[28] https://en.wikipedia.org/wiki/Quantum_entanglement
[29] https://people.cs.umass.edu/~strubell/doc/quantum_tutorial.pdf
[30] Giovannetti, V., Lloyd, S., Maccone, L. Quantum random access memory. Phys. Rev. Lett. 2008, 100, 160501. arXiv: 0708.1879
[31] https://peterwittek.com/understanding-quantum-svms.html
[32] Schuster, D. I., Sears, A. P., Ginossar, E., DiCarlo, F., Frunzio, L., Morton, J. J. L., Wu, H., Briggs, G. A. D., Buckley, B. B., Awschalom, D. D., & Schoelkopf, R. J. High-Cooperativity Coupling of Electron-Spin Ensembles to Superconducting Cavities. Phys. Rev. Lett. 2019, 105, 140501.
[33] Lloyd, S., Mohseni, M., Rebentrost, P. Quantum principal component analysis. 2013. arxiv.org/abs/1307.0401
[34] Lloyd, S., Mohseni, M., Rebentrost, P. Quantum principal component analysis. Nat. Phys. 2014, 10(631).
[35] Harrow, A. W., Hassidim, A., & Lloyd, S. Quantum Algorithm for Linear Systems of Equations. Phys. Rev. Lett. 2009, 103, 150502.
[36] Krenn, M. Automated Search for new Quantum Experiments. Physical Review Letters. PMID 26991161. 2016, 116(9),090405. arXiv:1509.02749. Bibcode:2016PhRvL.116i0405K. doi:10.1103/PhysRevLett.116.090405
[37] Wiebe, N., Braun, D., & Lloyd, S. Quantum Algorithm for Data Fitting. Physical Review Letters. PMID 23006156. 2012, 109(5), 050505. arXiv:1204.5242. Bibcode:2012PhRvL.109e0505W. doi:10.1103/PhysRevLett.109.050505
[38] Schuld, M., Sinayskiy, I., & Petruccione, F. Prediction by linear regression on a quantum computer. Physical Review A. 2016, 94(2), 022342. arXiv:1601.07823. Bib-code: 2016PhRvA.94b2342S. doi:10.1103/PhysRevA.94.02234
[39] Zhao, Z., Fitzsimons, J. K., & Fitzsimons, J. F. Quantum assisted Gaussian process regression. 2015. arXiv:1512.03929
[40] Berry, D. W., Childs, A. M., & Kothari, R. Hamiltonian simulation with nearly optimal dependence on all parameters. 56th Anual Symposium on Foundations of Computer Science. IEEE. 2015, 792–809. arXiv:1501.01715. doi:10.1109/FOCS.2015.54
[41] Aïmeur, E., Brassard, G., & Gambs, S. Quantum speed-up for unsupervised learning. Machine Learning. ISSN 0885-6125. 2013, 90(2),261–287. doi:10.1007/s10994-012-5316-5
[42] Wiebe, N., Kapoor, A., & and Svore, K. Quantum Algorithms for Nearest-Neighbor Methods for Supervised and Unsupervised Learning. Quantum Information & Computation. 2014, 15(3), 0318–0358. arXiv:1401.2142
[43] Wiebe, N., Kapoor, A., & Svore, K. M. Quantum Perceptron Models. Advances in Neural Information Processing Systems. 2016, 29, 3999–4007. arXiv:1602.04799
[44] https://www.nature.com/articles/s41534-018-0060-8
[45] Berry, D. W., Ahokas, G., Cleve, R., Sanders, B. C. Efficient quantum algorithms for simulating sparse Hamiltonians. Communications in Mathematical Physics. 2007, 270, 359–371.
[46] Adachi, S. H., Henderson, M. P. Application of quantum annealing to the training of deep neural networks. 2015. arXiv:1510.06356 [quant-ph]

[47] Denil, M., de Freitas, N. Toward the implementation of a quantum rbm. NIPS Deep Learning and Unsupervised Feature Learning Workshop. 2011, 5.

[48] Dumoulin V., Goodfellow, I. J., Courville, A., Bengio, Y. On the challenges of physical implementations of RBMs. 2013. arXiv:1312.5258 [stat.ML]

[49] Amin, M. H., Andriyash, E., Rolfe, J., Kulchytskyy, B., Melko, R. Quantum Boltzmann machine. 2016. arXiv:1601.02036 [quant-ph]

[50] Wadcock, J., Allen, E., Day, M., Frick, S., Hinchliff, J., Johnson, M., Morley-Short, S., Pallister, S., Price, A., & Stanisic, S. Advances in quantum machine learning. 2015. arXiv:1512.02900

[51] Neven, H., Denchev, V., Drew-Brook, M., Zhang, J., & Macready, W. NIPS 2009 Demonstration: Binary Classification using Hardware Implementation of Quantum Annealing. 2010.

[52] Neigovzen, R., Neves, J. L., Sollacher, R., & Glaser, S. J. Quantum pattern recognition with liquid-state nuclear magnetic resonance. Physical Review A. 2009, 79(4), 042321. arXiv:0802.1592. doi:10.1103/PhysRevA.79.042321

[53] Google Quantum A.I. Lab Team. Google Plus. 31 January 2017.

[54] NASA Quantum Artificial Intelligence Laboratory, NASA. 31 January 2017.

[55] Li, Z., Liu, X., Xu, N., & Du, J. Experimental Realization of a Quantum Support Vector Machine. Physical Review Letters. PMID 25910101. 2015, 114(14), 140504. arXiv:1410.1054, doi:10.1103/PhysRevLett.114.140504

[56] Tezak, N. & Mabuchi, H. A coherent perceptron for all-optical learning. EPJ Quantum Technology. 2015, 2. arXiv:1501.01608. doi:10.1140/epjqt/s40507-015-0023-3

[57] Li, Y., Holloway, G., Benjamin, S., Briggs, G. Baugh, J., & Mol, J. A simple and robust quantum memristor. Physical Review B. 2017, 96(7),075446. arXiv:1612.08409. doi:10.1103/PhysRevB.96.075446

[58] Benedetti, M., Realpe-Gómez, J., Biswas, R., & Perdomo-Ortiz, A. Quantum-assisted learning of graphical models with arbitrary pairwise connectivity. Physical Review X. 2017, 7(4), 041052. arXiv:1609.02542. doi:10.1103/PhysRevX.7.041052

[59] Yan, F., Iliyasu, A., & Le, P. Quantum image processing: A review of advances in its security technologies. International Journal of Quantum Information. 2017, 15(3), 1730001.

[60] Yao, X. W., Wang, H., Liao, Z., Chen, M. C., Pan, J., Li, J., Zhang, K., Lin, X., Wang, Z., Luo, Z., Zheng, W., Li, J., Zhao, M., Peng, X., & Suter, D. Quantum Image Processing and Its Application to Edge Detection: Theory and Experiment. Physical Review X. 2017, 7(3), 031041. arXiv:1801.01465

[61] https://ai.googleblog.com/2018/07/announcing-cirq-open-source-framework.html

[62] https://www.tensorflow.org/quantum

Alokananda Dey, Sandip Dey, Siddhartha Bhattacharyya,
Jan Platos, and Vaclav Snasel

5 Quantum inspired automatic clustering algorithms: A comparative study of Genetic algorithm and Bat algorithm

Abstract: This article is intendant to present two automatic clustering techniques of image datasets, based on quantum inspired framework with two different meta-heuristic algorithms, viz., Genetic Algorithm (GA) and Bat Algorithm (BA). This work provides two novel techniques to automatically find out the optimum clusters present in images and also provides a comparative study between the Quantum Inspired Genetic Algorithm (QIGA) and Quantum Inspired Bat Algorithm (QIBA). A comparison is also presented between these quantum inspired algorithms with their analogous classical counterparts. During the experiment, it was perceived that the quantum inspired techniques beat their classical techniques. The comparison was prepared based on the mean values of the fitness, standard deviation, standard error of the computed fitness of the cluster validity index and the optimal computational time. Finally, the supremacy of the algorithms was verified in terms of the p-value which was computed by *t-test* (statistical superiority test) and ranking of the proposed procedures was produced by the Friedman test. During the computation, the betterment of the fitness was judge by a well-known cluster validity index, named, DB index. The experiments were carried out on four Berkeley image and two real life grey scale images.

Keywords: automatic clustering, quantum computing, meta-heuristic algorithm, genetic algorithm, bat algorithm, DB Index, statistical test (*t-test*), Friedman test

5.1 Introduction

Clustering [1, 2] is a process of separating a heterogeneous dataset of different types of objects into some meaningful sets of homogeneous objects. Mathematically clustering can be described as follows. A dataset DS of K number of points can be represented as follows.

$DS = \{dp_1, dp_2, dp_3, \ldots, dp_K\}$, in which, each dp_i represents a single point of the specific dataset DS, where $i = \{1, 2, \ldots, K\}$. Now, in order to achieve a successful

Alokananda Dey, RCC Institute of Information Technology, Kolkata, West Bengal, India.
Siddhartha Bhattacharyya, CHRIST (Deemed to be University), Bangalore, Karnataka, India.
Sandip Dey, Sukanta Mahavidyalaya, Jalpaiguri, West Bengal, India.
Jan Platos, Vaclav Snasel, VSB Technical University of Ostrava, Czech Republic.

https://doi.org/10.1515/9783110670707-005

cluster CN of the dataset DS into p number of partition namely, $\{C_{N1}, C_{N2}, \ldots, C_{Np}\}$ the following properties must be hold.

$$C_{Ni} \neq \varnothing, \forall i = \{1, 2, \ldots, p\}.$$

$$C_{Ni} \cap C_{Nj} = \varnothing, \forall i, j = \{1, 2, \ldots, p\} \textit{ and } i \neq j.$$

$$\sum_{i=1}^{p} C_{Ni} = DS.$$

Clustering can be achieved in several ways and researchers are involved to design several methods for successful clustering, viz., K-means [3], Partitioning methods [4], Hierarchical clustering [5], Model-based clustering [6], Fuzzy clustering [7], Density-based clustering [8], etc. It is seen that among these methods, the K-means [3] out performs over others. In case of K-means algorithm an *apriori* knowledge is always required about the existing number of clusters within a dataset. On the other hand, an automatic clustering technique can be developed to discover the optimum number of clusters present in the dataset, in spite of knowing very less information or without having any information about the inspected dataset, which may yields usefulness in order to solve the purpose of segmentation as well as classification problem, which influenced the researchers to explore this field of clustering with automation.

Nature inspired meta-heuristic algorithms are very much useful for providing an optimal solution from a simple or complex optimization problems within a smallest time stamp. Some renowned nature inspired population based meta-heuristic algorithms are genetic algorithm (GA) [9], particle swarm optimization (PSO) [10], differential evolution (DE) [11], bacterial foraging optimization (BFO) [12], ant colony optimization (ACO) [13], bat algorithm [14, 15] etc. Some nature inspired meta-heuristic algorithms used for automatic clustering techniques are available in the literature [16–20]. Though, these algorithms can be able to identify a solution from a population within a minimum time frame still they may suffer from early convergence. In order to resolve the premature convergence problem, researchers are very much interested to associate the nature inspired meta-heuristic algorithms with the silent features of quantum computing, which yields the quantum inspired techniques to flourish day by day. Various kinds of computational problems can be solved by using the effects of different quantum mechanical phenomena which are presented in [21]. Several quantum inspired techniques associated with various meta-heuristic algorithms are invented till now and the efficiency of the quantum inspired techniques over their corresponding classical counterparts is established based on convergence, superior value of fitness and other related parameters [23–33]. These type of quantum inspired algorithms are developed for solving various kinds of optimization problems like mathematical function optimization [34], combinational optimization problem (knapsack problem) [35], multilevel image thresh holding [23–27], image analysis [22], automatic clustering [28–32] and task scheduling in distributed system [36], to name a few.

In this article, two different quantum inspired meta-heuristic algorithms, viz., QIGA and QIBA are proposed for automatic clustering technique of image datasets and also a comparison was produced between two different classical algorithms, viz., the classical genetic algorithm and the classical bat algorithm with their corresponding quantum influenced counterparts. The computed fitness values of the mean, standard deviation of that fitness, standard error of that fitness and the minimum computational time were the yardstick for adjudging the efficacy of these quantum inspired techniques over their analogous classical counterparts. Though, the experimental results are in favor of the quantum inspired techniques, still a statistical superiority test, viz., the unpaired *t-test* [37] was conducted and the corresponding evaluated p value establishes the efficacy of these quantum inspired techniques. Finally, Friedman test [38–39] was conducted among all the proposed procedure to judge the superiority of a procedure over another. The cluster validity index namely, DB index [40] was used as an objective function to compute the validity of the clustering.

The rest of the article is organized as follows. The overview of the genetic algorithm and the overview of the bat algorithm are summarized in Section 5.2 and 5.3 respectively. The fundamental idea of Quantum Computing is discussed in Section 5.4. The Section 5.5 presents the overview of the Cluster Validity Index (DB-Index). The steps involved in the proposed Quantum Inspired Genetic Algorithm and Quantum Inspired Bat Algorithm are presented in Section 5.6 and 5.7 respectively. The experimental outcomes and analysis are discussed in Section 5.8. Finally, a conclusion is drawn in Section 5.9.

5.2 Overview of Genetic algorithm

The well-known nature inspired meta-heuristic algorithm named, Genetic Algorithm (GA) is a kind of evolutionary algorithm based on the concept of Darwin's theory of evolution. John Holland is known as the inventor of GA, who proposed this algorithm in the year 1975 [9].

GA is a very useful population based adaptive searching process, in which each individual in a population is called a chromosome. In order to achieve the optimal solution GA basically explores the entire search space. During the search process the current population evolves to a same sized next population by applying three genetic operations, such as selection, crossover and mutation. In the successive iteration, the best chromosomes are selected among all the participating chromosomes from the previous pool and the selection is carried over depending upon their computed fitness value. Several methods can be used for the purpose of selection, such as Boltzmann selection, Roulette Wheel selection, Rank selection, Tournament selection etc. The crossover operation is used to produce a new pair of chromosomes by combining the portions of two or more randomly selected parent chromosomes which lead to produce a population by replacing the parent chromosomes by the new one. During the crossover operation one or multiple points can be selected as the partitioning points from each of the individual in the population. In order to ensure the genetic diversity in the population mutation operation takes place in which a randomly selected position of a randomly selected chromosome is altered depending upon a predefined value called mutation probability. This kind of alteration may be performed on a single point or multiple points of one or more randomly selected individuals in the population. In order to achieve faster convergence it is recommended to perform elitism at each generation. GA was already successfully applied in many applications like data clustering [17, 18], portfolio management [41], image processing [42] and many more.

Basic steps involved in Genetic algorithm are presented as follows.

1. Population of chromosomes is initialized randomly.
2. Fitness of each of the chromosome is evaluated.
3. Selection of chromosomes for next generation is done.
4. Do the following steps until the termination condition are satisfied.
5. Crossover is accomplished between a pair of randomly selected parent chromosomes to get child chromosomes, depending upon a predefined crossover probability.
6. Mutation is applied on the randomly selected offspring/ chromosome to introduce diversity in the population, depending upon a predefined mutation probability.
7. Selection is carried out for generating the next generation population.
8. End Loop.

5.3 Overview of Bat algorithm

In the year 2010, an efficient nature inspired population based meta-heuristic algorithm, viz., Bat algorithm and this algorithm was initially developed by Yang [14]. This algorithm is capable to identify the optimal solution or near optimal solution from a given problem, which can be an optimization problem of simple or complex type. Bats generally use echolocation for sensing the prey and recognizing their roosting crevices at the dark. They can also avoid their upcoming obstacles by using echolocation. Generally, bats can be able to produce a very loud sound pulse and listen for the bounced back echo from their surroundings. Depending upon the loudness of the response and the time delay from the emission to the reflection they can understand the shape, size and the velocity of the prey and can navigate accordingly. It is known that the high frequency emitted sound pulses influence them to give detailed information about their close environment so that they are able to detect the position of the prey precisely and allow them to move shorter distances.

But, if the emitted sound pulses have low frequency then they travel longer distance and provide rough information about their surroundings. Also the higher values of the loudness of the echolocation indicate that they are nearer to the prey and produce sound pulse very quickly. The following process describes the basic steps involved in the hunting strategy of the bats.

1. All bats are able to detect the distance between the prey/food and the obstacles of their surroundings by using the echolocation characteristics.
2. Bats are able to fly randomly from a position P_i with the velocity V_i, a fixed frequency F_{min}, with varying wavelength λ and loudness L_0 to search the prey.
3. Their loudness can be changed from a large number L_0 to a minimum constant number L_{min} and also the frequency can vary from F_{min} to F_{max}.

Let us considered in a search space, n numbers of bats are flying randomly. In a time stamp t, they are initiated their search from the position P_i having initial velocity V_i, frequency F_{min}, pulse rate r_i and loudness L_O. At time stamp $t + 1$, they may change their next position and velocity by using the following equations.

$$F_i = F_{min} + (F_{min} - F_{max})\beta \tag{5.1}$$

$$V_i^{t+1} = V_i^t + \left(P_i^{t+1} - P_{best}\right)F_i \tag{5.2}$$

$$P_i^{t+1} = P_i^t + V_i^{t+1} \tag{5.3}$$

where, $\beta \in [0, 1]$. The minimum and the maximum frequencies are $F_{min} = 0$ and $F_{max} = 2$ respectively. Here, P_{best} signifies the global best bat's location among the positions of the n numbers of bats in the current population. After evaluating the fitness of all solutions in the population, the position of the best solution among them is identified as the global best position. Now, it remains to check whether the bats are near or far away from the prey by using their pulse rates. A low pulse rate r_i with high probability ($rand_1 > r_i$) indicates that the bats will fly near to the current superior bat before carrying out a short random fly there. Otherwise, it is considered that the bats are nearer to the prey and they will do a random fly around their current position. Here, $rand_1$ is a uniformly distributed random number between [0, 1]. The following equation is used to represent the new positions of the bats.

$$P_{new} = P_{old} + \varepsilon L^t \tag{5.4}$$

where, $\varepsilon \in [-1, 1]$ and at the time stamp t, the average loudness of n bats is computed by $L^t = \frac{1}{n}\sum_{i=1}^{n} L_i^t$. After the random fly, if a bat is able to achieve a better position than the current global best position and its emitted sound is loud enough and also better than a random number ($rand_2 < L_i$), then it is considered that the bat will fly to this new position and the current global best position value will be updated with the newly generated position value. Here $rand_2$ is a random number which is distributed uniformly between [0, 1]. Now, the pulse rate (r_i) of the bats are increased and the loudness (L_i) of the bats are decreased simultaneously by the following equations.

$$r_i^{t+1} = r_i^0 \left(1 - e^{-\gamma t}\right) \tag{5.5}$$

$$L_i^{t+1} = \alpha L_i^t \tag{5.6}$$

where, $r_i^0 \in [0, 1]$ represents the initial pulse rate in which γ $(0 < \gamma < 1)$ and α $(0 < \alpha < 1)$ are constants. Now, the current P_{best} is selected by computing the fitness of all the bats.

Basic steps involved in Bat algorithm are presented as follows.

1. Bat population $\mathcal{P}$ is initialized randomly.
2. Pulse frequency ($\mathcal{F}$), velocity ($\mathcal{V}$), pulse rate (r_i) and loudness ($\mathcal{L}_0$) are initialized for each of the individual.
3. Fitness f_n of each individual is computed.
4. Do the following steps until the iteration number (J_n) reach to the specified maximum iteration number ($MaxJ$).
5. New bat population $\mathcal{P}_n$ is generated by adjusting the frequency ($\mathcal{F}_n$) and velocity ($\mathcal{V}_n$) of each individual. (eqs. (5.1)–(5.3)).
6. $if\ (rand > r_i) then$
7. Among all the best solutions, a single solution is selected and a local solution is produced around this selected solution (best solution). (eq. (5.4))
8. *end if*
9. By producing a random fly, a new solution is generated.
10. $if (rand < \mathcal{L}_{0i}\ and\ f_n(\mathcal{P}_i) < f_n(\mathcal{P}_{best})) then$
11. The new solution is accepted.
12. Pulse rate (r_i) is increased. (eq. (5.5))
13. Loudness ($\mathcal{L}_{0i}$) is decreased. (eq. (5.6))
14. *end if*
15. The current best bat is identified.
16. *end loop*

5.4 Fundamental idea of quantum computing

A classical computer is capable to hold a single binary bit 0 or 1 at a time. Unlike classical computer, a quantum computer is capable to store a unit of information which is called quantum bit or *qubit*, in which a single *qubit* is able to store the numbers zero (0) and one (1) at the same moment, which referred as the superposition state in quantum computing [24–26]. It is known that the *qubits* are able to accomplish numerous processes simultaneously which yields fastest execution of process. In a two-dimensional Hilbert space, a *qubit* are represented by a unit vector and are considered as a two state quantum-mechanical phenomena. In QC, the superpositions of the basis states are represented by the following equation.

$$|\Psi\rangle = \sum_{i=1}^{n} \mathcal{C}_i|\mathcal{V}_i\rangle$$

$$|\Psi\rangle = \mathcal{C}_1|\mathcal{V}_1\rangle + \mathcal{C}_2|\mathcal{V}_2\rangle + \ldots + \mathcal{C}_n|\mathcal{V}_n\rangle \tag{5.7}$$

where, $\mathcal{V}_i$ is the ith quantum states and $\mathcal{C}_i \in \mathbb{C}$. For a two states quantum bit or qubit, eq. (5.7) can be denoted by a linear superposition as $|\Psi\rangle = \mathcal{C}_1|0\rangle + \mathcal{C}_2|1\rangle$, in which, the state $|0\rangle$ is acknowledged as "ground state" and $|1\rangle$ is acknowledged as "excited state", and $\mathcal{C}_i$ belongs to the set of complex number which should satisfy the following normalization condition.

$$\sum_{i=1}^{n} |\mathcal{C}_i|^2 = 1 \tag{5.8}$$

In QC, there exist two striking features, viz., Coherence and Decoherence. Coherence is represented by the linear superposition ($|\Psi\rangle$) of the basis states whereas, decoherence is occurred when a forceful destruction is happened on the previously defined linear superposition. Quantum computing generally uses quantum logic gates as hardware devices.

A predefined unitary operator is used to update the *qubits* within a fixed time period. Mathematically, the unitary U operator quantum gates hold the relation $U^\dagger = U^{-1}$ and $UU^\dagger = U^\dagger U = I$.

In quantum computer a n qubit quantum register is able to store 2^n states simultaneously. As for example the eight states $|000\rangle$, $|001\rangle$, $|010\rangle$, $|011\rangle$, $|100\rangle$, $|101\rangle$, $|110\rangle$ and $|111\rangle$ require three quantum registers in which the three qubit system can be represented as $\begin{bmatrix} \alpha_1 & \alpha_2 & \alpha_3 \\ \beta_1 & \beta_2 & \beta_3 \end{bmatrix}$ where $\alpha_i^2 + \beta_i^2 = 1$, $i = 1, 2, 3$.

But in case of classical computer in order to store these eight states eight registers are required.

There are several quantum gates exists, such as Rotation Gate, Pauli – X Gate or NOT Gate, C-NOT Gate, Toffoli Gate, Fredkin Gate and Hadamard Gate [24–26] etc. and they can be used in several ways in different types of problems.

Rotation gate strategy can be used to update the values of (α_i, β_i) of the ith qubit by the following equation.

$$\begin{pmatrix} \alpha_i' \\ \beta_i' \end{pmatrix} = \begin{pmatrix} \cos\theta_i & -\sin\theta_i \\ \sin\theta_i & \cos\theta_i \end{pmatrix} \begin{pmatrix} \alpha_i \\ \beta_i \end{pmatrix} \tag{5.9}$$

Here, the rotation angle for each of the *qubit* is represented by θ_i, which is chosen according to specific problem. If α_i and β_i represent the probability amplitude of a *qubit* before rotation then after rotation they become α_i' and β_i' respectively and can be achieved by the following way.

$$\begin{pmatrix} \alpha_i' \\ \beta_i' \end{pmatrix} = \begin{pmatrix} \cos(\theta + \delta)_i \\ \sin(\theta + \delta)_i \end{pmatrix} \tag{5.10}$$

where $\begin{pmatrix} \alpha_i \\ \beta_i \end{pmatrix} = \begin{pmatrix} \cos \delta_i \\ \sin \delta_i \end{pmatrix}$

5.5 Cluster validity index

In the year 1979, David L. Davies and Donald W. Bouldin introduced an efficient cluster validity index, named, Davies-Bouldin (DB) Index [40], which was used in this article as an objective function to compute the fitness of each individual. The optimum number of clusters within a given dataset is determined by the calculated value of the *DB* index [40], where, minimum value of *DB* index [40] is taken for consideration. It can be considered as a function of the ratio between the sum of within-cluster scatter and the cluster separation. Let us consider the cluster similarity measure is CM_{ij} between two different clusters Cl_i and Cl_j. It is given by:

$$CM_{ij} = \frac{dM_i + dM_j}{d_{ci\_cj}} \tag{5.11}$$

Here dM_i be the dispersion measure of ith cluster and is denoted as

$$dM_i = \left[\frac{1}{tn\_o_i} \sum_{k=1}^{tn\_o_i} \left\| Dp_k^{(i)} - Cc_i \right\|^2 \right]^{\frac{1}{2}}$$

such that, tn_o_i denotes the total number of objects and Cc_i denotes the ith cluster center in cluster Cl_i and $\forall Dp_k^{(i)} \in Cl_i$.

The distance between the clusters Cl_i and Cl_j and can be defined as

$$d_{ci\_cj} = \left\| Cc_i - Cc_j \right\|.$$

The following equation is used to define the Davies – Bouldin (DB) index [40].

$$DB\_I = \frac{1}{no_c} \sum_{i=1}^{no_c} H_i \tag{5.12}$$

Where $H_i = \max_{j=1,2,\ldots,no_c,\ i \neq j} \left(CM_{ij} \right),\ i = 1, 2, \ldots, no_c$

5.6 Quantum inspired Genetic algorithm

5.6.1 Proposed methodology for QIGA

In this section, the operational strategy of quantum inspired GA algorithm is discussed. In order to recognize the optimum number of cluster presents within images, the proposed method was experimented on image dataset.

At first, a population (Pc) was initialized with N number of chromosomes. Each individual chromosome was of length Ln from the population was considered as a solution of the problem. (Line 1).

Now, the population (Pc) was encoded to produce the quantum states Q^s. The encoding process is discussed as follows.

Initially for each of the chromosome of the population a state $\theta^s_{i,j}\left(0 \le \theta^s_{i,j} \le 2\pi\right)$ was created randomly. Using the value of $\theta_{i,j}$ the quantum state Q^s was created. For a particular t^{th} generation the quantum state $Q^s(t)$ was represented as follows. $Q^s(t) = \{q^{st}_1, q^{st}_2, \ldots, q^{st}_N\}$,

where, $q^{st}_i (i = 1, 2, \ldots, N)$ denotes the arbitrary elements of $Q^s(t)$ and can be denoted as

$$q^{st}_i = \begin{bmatrix} \alpha^{st}_{i,1} \mid \alpha^{st}_{i,2} \mid \ldots \mid \alpha^{st}_{i,m} \mid \\ \beta^{st}_{i,1} \mid \beta^{st}_{i,2} \mid \ldots \mid \beta^{st}_{i,m} \mid \end{bmatrix}, \text{ here } m = Ln$$

and $\alpha^{st}_{i,j} = \cos\theta^{st}_{i,j}$ and $\beta^{st}_{i,j} = \sin\theta^{st}_{i,j}$ was computed from the random value of $\theta^{st}_{i,j}$, where $i = \{1, 2, \ldots, N\}$ *and* $j = \{1, 2, \ldots, Ln\}$.(Line 2).

Now, in order to generate the active cluster points a selection procedure was applied on the population and the selection process was guided by the state $Q^s(t)$ to produce a binary solution $Bc(t)$. The binary solution $Bc(t) = \{bc^t_1, bc^t_2, \ldots, bc^t_N\}$ was produced by observing the state $Q^s(t)$ in the initial generation $t = 1$.

Each of the binary solution bc^t_i was generated as follows.

$$bc^t_{i,j} = \begin{cases} 1, & If\ \beta^{st}_{i,j} > \alpha^{st}_{i,j} \\ 0, & Otherwise \end{cases}, \ s.t. i = \{1, 2, \ldots, N\}\ and\ j = \{1, 2, \ldots, Ln\}.$$

The criteria $bc^t_{i,j} = 1$ was set to identify the activated cluster points from the population $Pc(t)$ for the tth generation, where $i = \{1, 2, \ldots, N\}$ *and* $j = \{1, 2, \ldots, Ln\}$. (Line 3–8).

Now, the fitness of each individual chromosome from the quantum population was computed by an objective function (DB index [40] – eq. (5.12)).

The best chromosomes from the quantum population were chosen for generating the next generation population. (Line 9).

Now, a quantum rotation gate $Gs^t_{i,j}(\phi)$ was used for updating each quantum state $Q^s(t)$, which can be denoted as follows.

$$Gs_{i,j}^{t}(\phi) = \begin{bmatrix} \cos\phi_{i,j}^{st} & -\sin\phi_{i,j}^{st} \\ \sin\phi_{i,j}^{st} & \cos\phi_{i,j}^{st} \end{bmatrix},$$

where, the rotation angle $\phi_{i,j}^{st}$ was chosen randomly between (−0.5, 0.5).

The updated quantum state $Q^{new}(t)$ was identified by the eq. (5.10).(Line 11).

In case of quantum inspired genetic algorithm, the probability amplitude $(\alpha, \ \beta)$ of qubits were used to encode the chromosomes. After that quantum rotation gate was used to update $(\alpha, \ \beta)$, which lead to realize the genetic operation for generating the offspring instead of determining them from their parents as the chromosomes were in superposition or entangled state. In general, rotation influences the search space to extend as it is able to produce 2^n number of states from n qubit chromosome. The different values of $(\alpha, \ \beta)$ were responsible for generating the states (Lines 12–18).

A predefined mutation probability was used to perform the quantum mutation operation. The probability amplitude α would be converted to β and β would be converted to α for the greater value of the predefined mutation probability from a random number between (0,1). The following quantum NOT gate strategy was used to alter the probability amplitude $(\alpha, \ \beta)$ to $(\beta, \ \alpha)$.

$$\begin{bmatrix} 0 & 1 \\ 1 & 0 \end{bmatrix} \begin{bmatrix} \alpha_{i,j} \\ \beta_{i,j} \end{bmatrix} = \begin{bmatrix} \beta_{i,j} \\ \alpha_{i,j} \end{bmatrix}$$

$$\begin{bmatrix} 0 & 1 \\ 1 & 0 \end{bmatrix} \begin{bmatrix} \cos\theta_{i,j} \\ \sin\theta_{i,j} \end{bmatrix} = \begin{bmatrix} \sin\theta_{i,j} \\ \cos\theta_{i,j} \end{bmatrix} \text{ (Lines 19–26).}$$

The generation number t was updated by $t = t + 1$ and perform the above mentioned procedure for a predefined number of generation or until it met the stopping criteria. (Lines 10–27).

The best chromosome was identified and preserved along with its fitness and number of cluster.(Line 28).

The working principle for QIGA is presented as follows.

Algorithm: Quantum Inspired Genetic Algorithm for automatic clustering
Input Parameter:
Maximum number of generation: MxG
Size of population: N
Mutation probability: **μ**
Output parameter:
Optimum cluster number: $\mathcal{O\_N_C}$
Optimum fitness value: $\mathcal{O\_F_T}$

1. At the beginning, a population Pc was created randomly from an image dataset with N number of chromosomes where each chromosome was considered as a solution of the problem. During this process, the intensity value of each pixel of

the input image was normalized. The length of each chromosome was taken as Ln.The maximum pixel intensity value of the input image first passed through the square root function then was assigned to Ln. The image was also normalized between 0 & 1 and named it as I_{normal}.

2. Now quantum state $Q^P(t)$ was created and each chromosome of Pc was encoded to produce P^α and P^β using the information of the quantum state.
3. *for* $i=1$ *to N do*
4. *for* $j=1$ *to Ln do*
5. $\mathcal{N}_{Ci}$ numbes of active cluster centers were created from Pc_i by using the condition $P^\beta_{ij} > P^\alpha_{ij}$ and stored the cluster centers in CP_i.
6. Now, the fitness $\mathcal{F}_{Ti}$ was computed from CP_i with the help of a cluster validity index, named, *DB – Index*. [eq. (5.12)]
7. *end for*
8. *end for*
9. The best chromosome was identified and preserved along with its fitness value $\mathcal{F}_T \in \mathcal{F}_{Ti}, \forall i \in N$ and its corresponding number of cluster point $\mathcal{N}_C$.
10. *for* $t=1$ *to MxG do*
11. The quantum state $Q^N(t)$ was created by applying the rotation gate on $Q^P(t)$.
12. Continued step – 2 for the newly created quantum state $Q^N(t)$.
13. Continued steps- (3–8) for identifying the best chromosome.
14. *If* the best chromosome was identified in the population for the quantum state $Q^N(t)$ *then*
15. The value of the quantum state was updated by $Q^N(t)$
16. Otherwise
17. The previous value of the quantum state $Q^P(t)$ was preserved
18. *end if*
19. *for* $i=1$ *to N do*
20. *If* the current chromosome was not the best chromosome *then*
21. $j = rand(0,\ Ln)$
22. *If* $\mu > rand(0,1)$ *then*
23. Exchanged P^α_{ij} *and* P^β_{ij}
24. *end if*
25. *end if*
26. *end for*
27. *end for*
28. Finally, the algorithm produced and reported the optimum number of cluster $\mathcal{O}\_\mathcal{N}_C$ along with its associated value of fitness $\mathcal{O}\_\mathcal{F}_T$.

5.7 Quantum inspired Bat algorithm

5.7.1 Proposed methodology for QIBA

Now a day, the bat algorithm is widely applied in several types of complex optimization problems as it follows an easy mechanism for implementation and its quick convergence rate which sometimes leads a premature convergence. The premature convergence can be occurred for several reasons like the bats may be trapped into local optima due to the decreasing diversity of the bats or the most promising bat falls into the local optima then it influences all other bats to follow the same trajectory as other bats are instructed to follow the current optimal solution and moreover in bat algorithm there is no specified ways to get rid of the local optima. In order to solve this problem a quantum inspired bat algorithm was proposed in this article. The diversity of the bat population was achieved through quantum rotation gates which helped the proposed algorithm to avoid premature convergence without affecting the convergence speed though the efficiency of the proposed algorithm in terms of computational time was guaranteed [43].

The anticipated algorithm is established on the framework of bat algorithm with the flavor of c quantum computing. In order to control the exploitation and exploration, the parameters $\mathcal{BR}$ and $\mathcal{BA}$ are used and they are updated for guiding the local and global search process. Though, the new velocity and positions of each of the candidates are updated through the following equation.

$$V_i^{t+1} = \begin{cases} V_i^t, if(\delta_i, \delta_{best}) = (0,0) \\ V_i^t + \left(P_i^{t+1} - P_{best}\right), if(\delta_i, \delta_{best}) = (1,1) \\ V_i^t + \left(P_i^{t+1} - P_{best}\right) rand, if(\delta_i, \delta_{best}) = \{(1,0) or (0,1)\} \end{cases} \tag{5.13}$$

$$P_i^{t+1} = P_i^t + V_i^{t+1} \tag{5.14}$$

Where δ_i and δ_{best} represents the quantum state of the current bat and the local best bat respectively and *rand* is a random number between 0 and 1. If the state of both the current bat and the best bat are 0 it indicates that this current bat will not influence the search process during that point of time so there is no need to change the velocity of that position. If the state of both the current bat and the best bat are 1 it indicates that this current bat will very much influence the search process as it is close to the current optimal solution so it will change its current velocity using the information of the current optimal position. And if the state value of the current bat is just opposite of the state value of the best bat then it indicates that it is far away from the current optimal solution so it will like to make a random fly near the current best solution and update its velocity by using the information of the current optimal solution along with a random value between 0 and 1. Finally the new position of the current bat was computed by using the newly generated velocity of it.

During the search process the positions of each bat were updated as follows.

$$P_{new} = \begin{cases} P_{old} + \varepsilon AL^t & if\ rand > p\tau \\ P_{old} + \varepsilon(AL^t - P_{old}) & otherwise \end{cases} \tag{5.15}$$

Where, $\varepsilon \in [-1, 1]$ and at the time stamp t, $AL^t = \frac{1}{n}\sum_{i=1}^{n} AL_i^t$ represents the average loudness of n bats and $p\tau$ represents the pulse rate.

If the pulse rate $p\tau$ is low than a probability $(rand > p\tau)$ then it indicates that the bats will fly near to the current superior bat before carrying out a short random fly there. Otherwise, it is considered that the bats are nearer to the prey and they will do a random fly around their current position. Here *rand* is a uniformly distributed random number which is chosen between [0, 1] during the computation.

The pulse rate $(p\tau_i)$ is increased and the loudness (AL_i) of the bats is decreased by the following equations.

$$p\tau_i^{t+1} = p\tau_i^0\left(1 - e^{-\gamma t}\right) \tag{5.16}$$

$$AL_i^{t+1} = \alpha AL_i^t \tag{5.17}$$

Where, the constant term γ and the preliminary pulse rate $p\tau_i^0$ both were chosen between 0 and 1. Subsequently, the fitness values of every bats were estimated and the current P_{best} was selected. Here α was considered as a contraction – expansion and can be defined as follows.

$$\alpha = \alpha_i - \frac{\alpha_i - \alpha_\ell}{\mathcal{T}} t \tag{5.18}$$

Here α_i defines the initial value and α_ℓ defines the last value of α. The highest number of iteration is denoted by $\mathcal{T}$. During the execution $\alpha_i = 1$ and $\alpha_\ell = 0.5$ is set for increasing the performance of the algorithm [43, 44].

This section describes the process to identify the optimum number of clusters belong to the dataset of an image by the quantum inspired BAT algorithm.

Initially, the bat population P_BAT was initialized and considering the size of the population was *Psize*, in which each of the solution was of length *BLn* and each of the solution from the population was considered as a solution of the problem. The initialization was carried out with the normalized pixel intensity value of the input image, which was chosen randomly. The maximum intensity value of the input image first passed through the square root function then it was considered as the length of each particle (*BLn*). Another variable I_{normal} was used to store the normalized value (between 0 and 1) of each pixel intensity of the input image (Line 1).

Now, quantum state Q^{BAT} was created for each of the elements of P_BAT. And P_BAT was encoded by P_BAT^{α} and P_BAT^{β} the eq. (5.8) should be satisfied. The encoding process is shown as follows.

Initially, $\theta_{i,j}\left(0 \le \theta_{i,j} \le 2\pi\right)$ was created randomly for each of the particle of the solution of the population, which was used to produce the quantum state namely,

$Q^{\mathcal{B}AT}$. For a particular t^{th} generation the quantum state $Q^{\mathcal{B}AT}(t)$ was represented as follows.

$Q^{\mathcal{B}AT}(t) = \{q_1^{\mathcal{B}ATt}, q_2^{\mathcal{B}ATt}, \ldots, q_N^{\mathcal{B}ATt}\}$, where $q_i^{\mathcal{B}ATt}$ $(i = 1, 2, \ldots, Psize)$ denotes the arbitrary elements of $Q^{\mathcal{B}AT}(t)$ and can be represented as

$$q_i^{\mathcal{B}ATt} = \begin{bmatrix} \alpha_{i,1}^{\mathcal{B}ATt} & |\alpha_{i,2}^{\mathcal{B}ATt}| & \ldots & |\alpha_{i,m}^{\mathcal{B}ATt}| \\ \beta_{i,1}^{\mathcal{B}ATt} & |\beta_{i,2}^{\mathcal{B}ATt}| & \ldots & |\beta_{i,m}^{\mathcal{B}ATt}| \end{bmatrix}, \ (st.\ m = BLn)$$

here, each value of $\alpha_{i,j}^{st}$ and $\beta_{i,j}^{st}$ were computed by

$$\begin{pmatrix} \alpha_{i,j}^{\mathcal{B}ATt} \\ \beta_{i,j}^{\mathcal{B}ATt} \end{pmatrix} = \begin{pmatrix} \cos\theta_{i,j}^{\mathcal{B}ATt} \\ \sin\theta_{i,j}^{\mathcal{B}ATt} \end{pmatrix}, \text{ where } i = \{1, 2, \ldots, P\,size\} \text{ and } j = \{1, 2, \ldots, BLn\}$$

and the corresponding values of them are preserved in $P\_BAT_{i,j}^{\alpha}$ and $P\_BAT_{i,j}^{\beta}$ respectively. (Line 2).

The frequency $(\mathcal{BF}_i)$ and velocity $(\mathcal{BV}_i)$, Pulse Rate $(\mathcal{B}R_i)$ and Loudness $(\mathcal{B}A_i)$ were initialized for every quantum solutions inside the quantum population (Lines 3–4).

Now, in order to generate the active cluster points a selection procedure was applied on the population and the selection process was guided by the quantum state $Q^{\mathcal{B}AT}(t)$ to produce a binary solution $Bs(t)$. The binary solution $Bs(t) = \{bs_1^t, bs_2^t, \ldots, bs_N^t\}$ was produced by perceiving the state $Q^{\mathcal{B}AT}(t)$ in the initial generation $t = 1$.

Each of the binary solution bc_i^t was generated as follows.

$$bs_{i,j}^t = \begin{cases} 1, & If\, P\_BAT_{ij}^{\beta t} > P\_BAT_{ij}^{\alpha t} \\ 0, & Otherwise \end{cases}, \ st.i = \{1, 2, \ldots, P\,size\} \text{ and } j = \{1, 2, \ldots, BLn\}$$

The criteria $bs_{i,j}^t = 1$ was set to identify the activated cluster points from the population $P\_BAT\ (t)$ (Figure 5.1) for the tth generation, where $i = \{1, 2, \ldots, Psize\}$ and $j = \{1, 2, \ldots, BLn\}$. It was considered that the excited states are influenced to

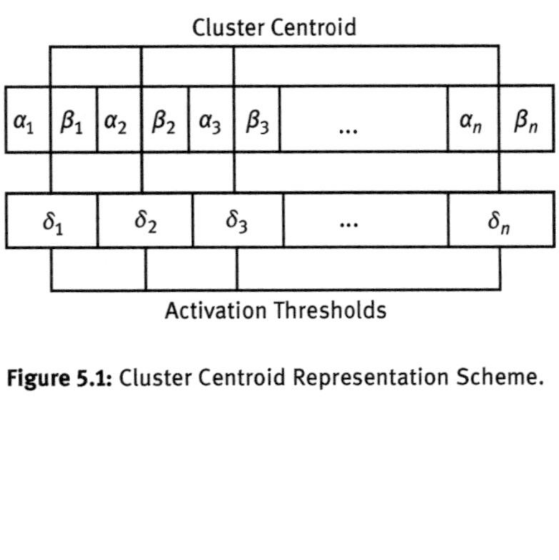

Figure 5.1: Cluster Centroid Representation Scheme.

make the active cluster centroids from each of the solutions of the population and are denoted by $\mathcal{N}_{Ci}$ for each of the solution of the population. The $\mathcal{N}_{Ci}$ number of activated unique cluster points was stored in $\mathcal{B}CP_i$.

Now, from $\mathcal{B}CP_i$, the fitness $(\mathcal{F}_{Ti})$ of each solution was evaluated with the help of a cluster validity index namely, *DB – Index*. [eq. (5.12)]. (Lines 5–10).

Now, the best fitness value $\mathcal{O}\_\mathcal{F}_{T}$ was identified from all computed fitness values $(\mathcal{O}\_\mathcal{F}_{Ti})$ and saved along with the respective number of cluster points $\mathcal{O}\_\mathcal{N}_{C}$. (Line 11).

Using rotation gate (eq. (5.10)) strategy another set of populations $P\_BAT^{\alpha}_{new}$ and $P\_BAT^{\beta}_{new}$ were created. Update the values of P_BAT^{α} and P_BAT^{β} by using their earlier values or by using their recently created values of $P\_BAT^{\alpha}_{new}$ and $P\_BAT^{\beta}_{new}$ on the basis of the best value of fitness. Now, if the recently created values were not able to make better fitness values then the current set of values were rejected and the old set of values were preserved in the population otherwise the new set of values were accept. Now, it was possible to create a novel quantum population as it produces the diversity in the solution space (Line 12).

The rotation gate updating process was applied which yielded the best fitness value $(\mathcal{B}\mathcal{F}_{T})$ and the respective cluster points were preserved in $\mathcal{B}CP$ and $\mathcal{B}nc$ conserved the total number of cluster points (Line 13).

During each execution of the main loop, the quantum positions and velocities were updated by the eqs. (5.13) and (5.14) (Line 16).

The rotation gate strategy was responsible for the improvement of the best quantum solutions (Line 17).

The best solution was selected from the computed fitness of the updated positions and a local solution was generated by using the eq. (5.15). (Lines 18–20).

Next, the new quantum solutions were estimated and by applying rotation gate strategy each value of the state $Q^{\mathcal{B}AT}(t)_J$, $P\_BAT^{\alpha}_J$ and $P\_BAT^{\beta}_J$ were updated. (Lines 21–22).

Now the new solution was accepted and the pulse rate of each bat was increased and the loudness of each bat was decreased by the eqs. (5.16) and (5.17). (Lines 23–26).

The best quantum solutions was determined and saved (Line 27).

In order to achieve the best solution the main loop was continued for a predefined number of times (Lines 14–29).

Finally, the best quantum solution was determined and the optimum number of cluster was identified along with its associated value of fitness. (Line 30).

The working principle for QIBA is presented as follows.

Algorithm: Quantum Inspired Bat Algorithm for automatic clustering
Input parameter:
Maximum number of Iteration: *MI*
Size of population: *Psize*

Output parameter:
Optimum cluster number: $\mathcal{O}\_\mathcal{N}_\mathcal{C}$
Optimum fitness value: $\mathcal{O}\_\mathcal{F}_\mathcal{T}$

1. A population P_BAT having $Psize$ number of particles (bat) was initialized.
2. Quantum state $Q^{\mathcal{B}AT}(t)$ was created for each of the elements of P_BAT. And in order to encode P_BAT the information of the quantum state was used and produce P_BAT^α and P_BAT^β.
3. The frequency $(\mathcal{BF})_i$ and velocity $(\mathcal{BV}_i)$ were initialized for every solution of the population.
4. The loudness $(\mathcal{B}A_i)$ and the pulse rates $(\mathcal{B}R_i)$ were also initialized.
5. *for* $i = 1$ *to Psize do*
6. *for* $j = 1$ *to BLn do*
7. $\mathcal{N}_{\mathcal{C}i}$ numbers of cluster points from P_BAT_i were created by satisfying the condition $P\_BAT_{ij}^\beta > P\_BAT_{ij}^\alpha$ and the cluster points were stored in $\mathcal{B}CP_i$.
8. Now, the fitness $\mathcal{F}_{\mathcal{T}i}$ was evaluated from $\mathcal{B}CP_i$ by using the cluster validity index $(DB - Index)$.
9. *end for.*
10. *end for.*
11. The best solution in terms of the best fitness value $\mathcal{O}\_\mathcal{F}_\mathcal{T}$ was identified.
12. Now, the quantum state $Q^{\mathcal{B}AT}(t)$ was updated by using quantum rotation gate to generate new set of values of P_BAT^α and P_BAT^β.
13. The steps 5 to 12 were continued until a predefined condition was satisfied in order to identify the best fitness value $\mathcal{BF}_\mathcal{T}$.
14. *for* $I = 1$ *to MI do*
15. *for* $J = 1$ *to Psize do*
16. The velocity and location of every particles in P_BAT_J were modified by using the eqs. (5.13) and (5.14).
17. Each value of the state $Q^{\mathcal{B}AT}(t)_J$, $P\_BAT_J^\alpha$ and $P\_BAT_J^\beta$ were updated using the rotation gate strategy in order to achieve a better solution from the population.
18. *if* $(Rand\ (0,1) > \mathcal{B}R_{ij})$ then
19. The best solution was selected after calculating the fitness of all bats from their updated positions and a native solution was generated with the help of the eq. (5.15).
20. *end if.*
21. Now, new solution was evaluated.
22. Each value of the state $Q^{\mathcal{B}AT}(t)_J$, $P\_BAT_J^\alpha$ and $P\_BAT_J^\beta$ were updated by using the rotation gate strategy in order to achieve a better solution from the population.
23. *if* $(Rand\ (0,1) < \mathcal{B}A_i\ \&\mathcal{BF}_{\mathcal{T}new} < \mathcal{BF}_{\mathcal{T}Best})$ then
24. New solution was accepted.
25. $\mathcal{B}R_i$ was increased and $\mathcal{B}A_i$ was decreased by using the eqs. (5.16) and (5.17).
26. *end if.*

27. The best solution was chosen.
28. *end for.*
29. *end for*
30. Finally, the algorithm produced and reported the optimal number of cluster points $\mathcal{O}\_\mathcal{N}_\mathcal{C}$ with its associated fitness value $\mathcal{O}\_\mathcal{F}_\mathcal{T}$.

5.8 Experiment and result analysis

The aim of this article is to propose two different quantum inspired algorithms combined with the working procedure of two dissimilar meta heuristic algorithms namely, Genetic algorithm and Bat algorithm, which were accomplished to recognize the optimum number of clusters present within a given image. In order to accomplish the experiment, the DB – Index [40] was used as an objective function. The optimal result was realized for the minimum value obtained from the computation. The following subsections presents the results obtained from the experiment.

5.8.1 Representation scheme of cluster centroids

The cluster centroids within a given image dataset were used to automatically recognize the optimal number of cluster points present in that dataset. During the execution of the process, the following equation was used to choose the cluster centroids.

$$\delta_i = \begin{cases} 1 \text{ } if \text{ } \beta_i > \alpha_i \\ 0 \text{ } Otherwise \end{cases} \tag{5.19}$$

where, $\delta_i(1 < i < n)$ was the activation threshold. By observing the value of δ_i, the cluster centroids were chosen from a single solution. Here α_i and β_i were the complex coefficients and δ_i was considered similar to the state of the quantum bit. A value of 1 for δ_i represents the activated cluster center implies that the quantum bit is in excited state otherwise it is in the ground state. The following Figure 5.1 is showing the Cluster Centroid representation scheme.

5.8.2 Experimental result analysis

In this article two different population based quantum inspired meta-heuristic algorithms, viz., quantum inspired genetic algorithm and quantum inspired bat algorithm are presented. The experiment was performed over four Berkeley images and two real life images. The performance of any meta-heuristic algorithm is very much dependent upon the settings of the parameter i.e., an appropriate selection of parameters can influence the convergence speed of the meta-heuristic algorithm to a great level.

The parameters for QIGA were chosen as Number of generations was considered for the range of values 100 to 500; Population size was taken from 25 to 50. Mutation probability was chosen as 0.01. And for CGA the number of generations was considered from 100 to 500. Population size was taken from 25 to 50. Crossover and Mutation probability was chosen as 0.9 and 0.01 respectively. The parameters for QIBA and CBA were chosen as number of iterations from 100 to 300, Population size was taken from 25 to 50, $\gamma = \alpha = 0.5$, initially pulse rate could be denoted as $r_i^0 \in [0, 1]$, the wavelength range was $[\lambda_{min}, \lambda_{max}] = [0, 1]$, and the range of frequency was $[F_{min}, F_{max}] = [0, 1]$ and the loudness $L_0 = 0.5$.

During the Experiment it was seen that the quantum inspired methods can produce a good result without having a large population or large number of iteration than their corresponding classical methods. Due to the probability amplitude (α, β) of qubits a single solution may generate several solutions which may lead to increase the size of a small population. So in case of quantum inspired methods the search can start with a small population. For each of the specific settings of parameters each procedure was run for 30 times. The mean value of the fitness was computed from several runs. The computed valued from several runs was presented in the following tables.

The optimal number of clusters along with their associated optimal fitness value for all the implemented algorithms like QIGA, CGA, QIBA and CBA were presented in Table 5.1. From Table 5.1 it can be stated that the computed fitness value from the method QIBA was providing the best fitness value (minimum value) among all other implemented methods for every image datasets and it was also seen that the better fitness values were achieved from quantum inspired methods than their classical counterparts. In Table 5.2 the optimal execution time for all the

Table 5.1: Optimal value of the cluster number ($\mathcal{O\_N}_\mathcal{C}$), Optimal fitness value ($\mathcal{O\_F}_\mathcal{T}$) for QIGA, CGA, QIBA and CBA.

Dataset	QIGA		CGA		QIBA		CBA	
	$\mathcal{O\_N}_\mathcal{C}$	$\mathcal{O\_F}_\mathcal{T}$	$\mathcal{O\_N}_\mathcal{C}$	$\mathcal{O\_F}_\mathcal{T}$	$\mathcal{O\_N}_\mathcal{C}$	$\mathcal{O\_F}_\mathcal{T}$	$\mathcal{O\_N}_\mathcal{C}$	$\mathcal{O\_F}_\mathcal{T}$
#86000	4	0.241241	3	0.310025	5	0.202196	5	0.239741
#92059	4	0.225393	4	0.297862	4	0.191752	4	0.226332
#94079	4	0.250224	5	0.319232	4	0.222669	5	0.241601
#97017	5	0.307631	3	0.350092	5	0.300434	5	0.337940
couple	5	0.146938	5	0.197734	5	0.119231	5	0.148203
clown	6	0.244835	5	0.244303	6	0.185826	6	0.209202

Table 5.2: Optimal execution time ($\mathcal{O\_T_E}$) in second for QIGA, CGA, QIBA and CBA.

Dataset	QIGA ($\mathcal{O\_T_E}$)	CGA ($\mathcal{O\_T_E}$)	QIBA ($\mathcal{O\_T_E}$)	CBA ($\mathcal{O\_T_E}$)
#86000	49.83	74.17	12.66	38.91
#92059	55.56	102.89	22.36	63.87
#94079	53.81	97.95	21.18	49.77
#97017	56.22	111.31	17.45	40.33
couple	123.09	238.87	68.23	97.29
clown	130.11	251.69	72.37	95.82

implemented methods was presented which shows that the quantum inspired methods were taking less time than their classical counterparts and the processing time taken by QIBA was the minimal among other methods. Table 5.3 provides the computed mean value, standard deviation and standard error from QIGA and CGA. Here it is seen that QIGA is providing more promising results than CGA in terms of mean, standard deviation and standard error. Similarly QIBA is providing more promising results than CBA which is presented in Table 5.4. From these two tables it is seen that QIBA has the minimum error among all other. The statistical superiority test named unpaired *t- test* was performed between QIGA and CGA and between QIBA and CBA which is presented in Table 5.5. From Table 5.5 it is seen that out of six results three results are extremely significant and remaining three results are very significant for the comparison between QIGA and CGA. And for QIBA and CBA, out of six results four results are extremely significant, one result is very significant and one result is significant. This analysis indicates that the quantum inspired

Table 5.3: Mean value of fitness ($\mathcal{F}_\mu$), Standard deviation of fitness ($\mathcal{F}_\sigma$) and Standard error of fitness ($\mathcal{S_E}$) for QIGA, CGA.

Dataset	QIGA			CGA		
	$\mathcal{F}_\mu$	$\mathcal{F}_\sigma$	$\mathcal{S_E}$	$\mathcal{F}_\mu$	$\mathcal{F}_\sigma$	$\mathcal{S_E}$
#86000	0.259923	0.031113	0.007723	0.318324	0.061119	0.102001
#92059	0.248769	0.043454	0.009898	0.301971	0.054612	0.009933
#94079	0.280010	0.048452	0.009336	0.335100	0.052379	0.006150
#97017	0.322167	0.062244	0.008253	0.374516	0.074216	0.012001
couple	0.156938	0.056512	0.012636	0.214296	0.069995	0.017591
clown	0.259843	0.034047	0.009269	0.276215	0.045183	0.013615

Table 5.4: Mean value of fitness ($\mathcal{F}_\mu$), Standard deviation of fitness ($\mathcal{F}_\sigma$) and Standard error of fitness ($\mathcal{S}_\mathcal{E}$) for QIBA and CBA.

Dataset	QIBA			CBA		
	$\mathcal{F}_\mu$	$\mathcal{F}_\sigma$	$\mathcal{S}_\mathcal{E}$	$\mathcal{F}_\mu$	$\mathcal{F}_\sigma$	$\mathcal{S}_\mathcal{E}$
#86000	0.217982	0.022745	0.006546	0.287976	0.027456	0.007126
#92059	0.221745	0.025513	0.008571	0.293700	0.036889	0.009193
#94079	0.228774	0.045474	0.007238	0.277012	0.064098	0.010117
#97017	0.315692	0.038725	0.007321	0.371454	0.047451	0.009676
couple	0.116231	0.043081	0.009347	0.165203	0.048741	0.014862
clown	0.199537	0.024239	0.005364	0.214552	0.025914	0.006518

Table 5.5: Result of unpaired $t-test$ ($\mathcal{P}-value$) between QIGA and CGA and between QIBA and CBA.

Dataset	QIGA & CGA		QIBA & CBA	
	$\mathcal{P}$ – value	Significant Level	$\mathcal{P}$ – *value*	Significant Level
#86000	<0.0001	1	<0.0001	1
#92059	0.0002	2	0.0451	3
#94079	0.0001	2	0.0001	2
#97017	<0.0001	1	<0.0001	1
couple	0.0016	2	<0.0001	1
clown	<0.0001	1	<0.0001	1

Level of Significance: 1 – Extremely Significant, 2 – Very Significant, 3 – Significant

methods are much more superior to their classical counterparts. Finally, the result of Friedman test for QIGA, CGA, QIBA and CBA is presented in Table 5.6. As per the ranking QIBA is being the most preferable method as it has achieved the first rank.

5.8.3 Simulation of work

Python environment was used to carry out the experimental process. The experiment was done on four Berkeley images, two of dimensions 80 × 120 and other two of dimensions 120 × 80 and two real life images of dimensions 512 × 512. The development environment was consisting of Windows 7 with the configuration facility of DELL Intel(R) Core(TM) i3, 2.00 GHz and 4.00 GB RAM.

The efficiency of the suggested methods over their classical counterparts was proved depending upon computed mean fitness value, standard deviation, standard error and smallest computation time. Moreover, the competence of the quantum inspired algorithms over their classical algorithms was justified by performing the $t - test$ (statistical superiority test). The optimal number of cluster ($\mathcal{O}\_\mathcal{N}_\mathcal{C}$) and its associated optimal fitness value ($\mathcal{O}\_\mathcal{F}_\mathcal{T}$) for all the algorithms (QIGA, CGA, QIBA and CBA) were reported in Table 5.1. The optimal execution time ($\mathcal{O}\_\mathcal{T}_\mathcal{E}$) in second for all the algorithms (QIGA, CGA, QIBA and CBA) were shown in Table 5.2. The mean fitness value ($\mathcal{F}_\mu$), standard deviation ($\mathcal{F}_\sigma$) and standard error ($\mathcal{S}_\mathcal{E}$) of the algorithms QIGA and CGA and were demonstrated in Table 5.3 and the same things for the algorithms QIBA and CBA were demonstrated in Table 5.4. Finally, the result of $t - test$ [37] was presented in Table 5.5 which was performed with 95% confidence level. During the test, the $\mathcal{P} - value$ is checked. If it is less than 0.05 then null hypotheses is considered for rejection against the alternative hypothesis. Finally, Friedman test [38, 39] was performed among the participating procedure and was enlisted in Table 5.6. The obtained rank from the proposed algorithms QIGA, CGA, QIBA and CBA are 2.66, 3.91, 1.08 and 2.33 respectively. According to the rank value, QIBA can be considered as the superior algorithm as it has the minimum rank value.

Table 5.6: Result of Friedman test for QIGA, CGA, QIBA and CBA. The rank for each procedure is shown within the parenthesis.

Dataset	QIGA	CGA	QIBA	CBA
#86000	0.24 (3)	0.31 (4)	0.20 (1)	0.23 (2)
#92059	0.22 (2.5)	0.29 (4)	0.19 (1)	0.22 (2.5)
#94079	0.25 (3)	0.31 (4)	0.20 (1)	0.24 (2)
#97017	0.30 (1.5)	0.35 (4)	0.30 (1.5)	0.33 (3)
couple	0.14 (2.5)	0.19 (4)	0.10 (1)	0.14 (2.5)
clown	0.24 (3.5)	0.24 (3.5)	0.18 (1)	0.20 (2)
Average Rank	2.66	3.91	1.08	2.33

5.8.4 Dataset used

The following images (Figure 5.2) were used during the experiment and during the accomplishment of the process the normalized values between (0,1) were considered for the images.

Figure 5.2: Test Images: (a) #86000 of size 80 × 120, (b) #92059 of size 80 × 120, (c) #94079 of size 120 × 80, (d) #97017 of size 120 × 80, (e) couple of size 512 × 512, (f) clown of size 512 × 512.

5.9 Conclusion

This article is envisaged an automatic clustering technique to recognize the optimal number of cluster points present in an image dataset, which shows a new way to combine the features of quantum computing with two different meta-heuristic algorithms, viz., Genetic algorithm and Bat algorithm. These proposed techniques were found superior over their classical counterparts as they out performed over corresponding classical counterparts for automatically identifying the optimal number of cluster points present in an image dataset within a significantly small time frame. Also the effectiveness of the quantum inspired algorithms were judged and presented depending upon the computed value of optimal fitness, mean fitness value, standard deviation of the fitness, standard error of the fitness and more over the computed $p-value$ from the statistical superiority test $t-test$. Finally, Friedman test were performed to identify the superior technique between the quantum inspired versions with their classical version. Moreover, it was seen that the quantum inspired Bat algorithm was the best performing procedure among the others. All the computed results were presented in the previous tables in order to proof the effectiveness of the quantum inspired methods over their classical counterparts as well as prove the superiority of QIBA. So far, only gray level images were used for experimental purpose and the proposed techniques were developed to satisfy only one objective at a time.

Currently authors are involved to explore the realm of quantum computing to resolve the concerned multi objective optimization problems effectively and efficiently within a smaller time frame and also trying to take care of the true colour images for this direction of the research.

References

[1] Jain, A., & Dubes, R. Algorithms for Clustering Data. Prentice Hall, 1988.

[2] Jain, A. K., Murty, M. N., & Flynn, P. J. Data clustering: A review. ACM Computing Surveys. 1999, 31(3),264–323.

[3] Yadav, J. & Sharma, M. A Review of K-mean Algorithm. International Journal of Engineering Trends and Technology (IJETT). July 2013, 4(7).

[4] Saket, S., Pandya, J, S. An Overview of Partitioning Algorithms in Clustering Techniques. International Journal of Advanced Research in Computer Engineering & Technology (IJARCET). June 2016, 5(6).

[5] Pestunov, I., Rylov, S. A., & Berikov, V. Hierarchical clustering algorithms for segmentation of multispectral images. Optoelectronics, Instrumentation and Data Processing. 2015, 51, 329–338.

[6 Wehrens, R., Buydens, L., Fraley, C., & Raftery, A. Model-Based Clustering for Image Segmentation and Large Datasets via Sampling. Journal of Classification. 2004, 21(2), 231–253.

[7] Chen, M. & Ludwig, S. A. Color Image Segmentation Using Fuzzy C-Regression Model. Hindawi Advances in Fuzzy Systems. 2017, 4582948, 15.

[8] Ye, Q., Gao, W, & Zeng, W. Color image segmentation using density-based clustering. 2003 International Conference on Multimedia and Expo. ICME '03. Proceedings. Baltimore, MD, USA, 2003, 03TH8698, II–401.

[9] Holland, J. Adaptation in Natural and Artificial Systems: An Introductory Analysis with Applications to Biology, Control and Artificial Intelligence. The University of Michigan Press, August 1975, 53.

[10] Hui-liana, F. & and Xian-lib, L. Discrete particle swarm optimization for TSP based on neighborhood. Application Research of Computers. 2011, 2, 030.

[11] Storn, R. & Price, K. Differential evolution – a simple and efficient heuristic for global optimization over continuous spaces. Technical Report TR-95-012, ICSI, 1995.

[12] Passino, K. M. Bacterial foraging optimization. Innovations and Devel-opments of Swarm Intelligence Applications. 2012, 219–233.

[13] Odili, J. B., Deng, H., Yang, C., Sun, Y., et al. Application of Ant Colony Optimization to Solving the Traveling Salesman's Problem. Science Journal of Electrical & Electronic Engineering. 2013, 2013.

[14] Yang, X.-S. A new metaheuristic bat-inspired algorithm. Studies in Computational Intelligence. 2010, 284, 65–74.

[15] Alihodzic, A. & Tuba, M. Improved hybridized bat algorithm for global numerical optimization. Proceedings of the 16th IEEE International Conference on Computer Modelling and Simulation (UKSim-AMSS '14). March 2014, 57–62.
[16] Liuabcd, Y., Wud, X., Shenb, Y. Automatic clustering using genetic algorithms. Applied Mathematics and Computation, 15 October 2011, 218 (4), 1267–1279.
[17] Raposo, C., Antunes, C. H., Barreto, J. P. Automatic Clustering Using a Genetic Algorithm with New Solution Encoding and Operators. Automatic Clustering Using a Genetic Algorithm with New Solution Encoding and Operators. Murgante, B. et al. (Eds.). Computational Science and Its Applications – ICCSA 2014. Lecture Notes in Computer Science, Springer, 2014, 8580.
[18] Garai, G., Chaudhuri, B. B. A novel genetic algorithm for automatic clustering. Journal Pattern Recognition Letters Archive. Elsevier Science Inc. New York, NY, USA, 19 January 2004, 25(2), 173–187.
[19] Das, S., Ajith, A., & Konar, A. Automatic Clustering Using an Improved Differential Evolution Algorithm. Systems, Man and Cybernetics, Part A: Systems and Humans, IEEE Transactions, 2008, 38, 218–237. 10.1109/TSMCA.2007.909595
[20] Abubaker, A., Baharum, A., Alrefaei, M. Automatic Clustering Using Multi-objective Particle Swarm and Simulated Annealing. 1 July 2015, 10(7), e0130995. doi: 10.1371/journal.pone.0130995
[21] McMohan, D. Quantum Computing Explained. John Wiley & Sons, Inc., Hoboken, New Jersey, 2008.
[22] Dey, S., Bhattacharyya, S., & Maullik, U. Quantum Behaved Swarm Intelligent Techniques for Image Analysis: A Detailed Survey. Handbook of Research on Swarm Intelligence in Engineering. 2015.
[23] Dey, S., Saha, I., Bhattacharyya, S., & Maulik, U. Multi-level thresholding using quantum inspired meta-heuristics. Knowledge-Based Systems. 2014, 67, 373–400.
[24] Dey, S., Bhattacharyya, S., & Maullik, U. Optimum Gray Level Image Thresholding using a Quantum Inspired Genetic Algorithm. Advanced Research on Hybrid Intelligent Techniques and Applications. 2015, 201.
[25] Dey, S., Bhattacharyya, S., & Maulik, U. Quantum inspired genetic algorithm and particle swarm optimization using chaotic map model based interference for gray level image thresholding. Swarm and Evolutionary Computation. 2014, 15, 38–57.
[26] Dey, S., Bhattacharyya, S., & Maullik, U. Efficient quantum inspired meta-heuristics for multi-level true color image thresholding. Applied Soft Computing. 2017, 56, 472–513.
[27] Dey, S., Bhattacharyya, S., & Maullik, U. New quantum inspired meta-heuristic techniques for multi-level color image thresholding. Applied Soft Computing. 2016, 46, 677–702.
[28] Dey, S., Bhattacharyya, S., & Maullik, U. Quantum inspired automatic clustering for multi-level image thresholding. Proceedings of International Conference On Computational Intelligence and Communication Networks (ICCICN 2014), RCCIIT, Kolkata, India. 2014, 247–251.
[29] Dey, S., Bhattacharyya, S., & Maullik, U. Quantum inspired automatic clustering for multi-level image thresholding. Proceedings of International Conference On Computational Intelligence and Communication Networks (ICCICN 2014), RCCIIT, Kolkata, India. 2014, 242–246
[30] Dey, S., Bhattacharyya, S., Snasel, V., Dey, A., & Sarkar, S. PSO and DE based novel quantum inspired automatic clustering techniques. 2017, 285–290. 10.1109/ICRCICN.2017.8234522
[31] Dey, A., Dey, S., Bhattacharyya, S., Snasel, V., & Hassanien, A. E. Simulated Annealing Based Quantum Inspired Automatic Clustering Technique. Hassanien, A., Tolba, M., Elhoseny, M., Mostafa, M. (Eds.). The International Conference on Advanced Machine Learning Technologies and Applications (AMLTA2018). Advances in Intelligent Systems and Computing. Springer, Cham, 2018, 723.

[32] Dey, S., Bhattacharyya, S., & Maullik, U. Quantum-Inspired Automatic Clustering Technique Using Ant Colony Optimization Algorithm. Quantum-Inspired Intelligent Systems for Multimedia Data Analysis. 2018. 10.4018/978-1-5225-5219-2.ch002
[33] Narayanan, A. & Moore, M. Quantum-inspired genetic algorithms. Proc. IEEE Evolutionary Computation. 1996, 61–66.
[34] Yang, Y., Kuo, S., Lin, F., Liu, I., & Chou, Y. Improved Quantum-Inspired Tabu Search Algorithm for Solving Function Optimization Problem. 2013 IEEE International Conference on Systems, Man, and Cybernetics, Manchester. 2013, 823–828.
[35] Han, K.-H., & Kim, J.-H. Quantum-inspired evolutionary algorithm for a class of combinatorial optimization. IEEE Transactions on Evolutionary Computation. December 2002, 6(6),580–593.
[36] Alam, T., Gandhi, T., & Nitin, N. Quantum Genetic Algorithm with Rotation Angle Refinement for Dependent Task Scheduling on Distributed Systems. 2017. 10.1109/IC3.2017.8284295
[37] Flury, B. A First Course in Multivariate Statistics. Springer-Verlag, Berlin, Germany, 1997.
[38] Friedman, M. The use of ranks to avoid the assumption of normality implicit in the analysis of variance. Journal of the American Statistical Association. 1937, 32, 674–701.
[39] Friedman, M. A comparison of alternative tests of significance for the problem of m rankings. Annals of Mathematical Statistics. 1940, 11, 86–92.
[40] Maulik, U., & Bandyopadhyay, S. Performance evaluation of some clustering algorithms and validity indices. IEEE PAMI. 2002, 24, 1650–1654.
[41] Aranha, C. Portfolio management by genetic algorithm with error modeling. Information Sciences. 2007, 459–465.
[42] Paulinas, M., & Usinskas, A. A survey of genetic algorithms applications for image enhancement and segmentation. Information Technology and Control. 2007, 36, 278–284.
[43] Dey, A., Bhattacharyya, S., Dey, S., Platos, J., & Snasel, V. Quantum-Inspired Bat Optimization Algorithm for Automatic Clustering of Grayscale Images. Recent Trends in Signal and Image Processing. 2019, 89–101. 10.1007/978-981-13-6783-0_9
[44] Sun, J., Fang, W., Wu, X., Palade, V., & Xu, W. Quantum-Behaved Particle Swarm Optimization: Analysis of Individual Particle Behavior and Parameter Selection. Evolutionary Computation. September 2012, 20(3),349–393.

Siddhartha Bhattacharyya

6 Conclusion

Machine learning is a branch of Artificial intelligence, which tries to develop techniques that have ability to learn from experiences. These experiences are articulated as data generated from underlying models. The goal of any machine learning technique is to abstract the underlying model reliably for specific tasks like prediction and decision making. Enticed by the versatility of quantum mechanics, scientists have embarked on evolving machine learning algorithms based on the principles of quantum mechanics. As a result, different incarnations of the classical machine learning algorithms have come up enveloped within the quantum computing mechanisms. This book intends to throw a light on some of the recent advancements in this direction.

The book first introduces to the readers the basics of quantum machine learning with reference to the well-known quantum algorithms in the form of Grover's search algorithm [1], quantum reinforcement learning [2–6] and quantum annealing [7, 8]. A bird's eye view on the evolution of quantum neural networks [9–17] is also presented. The information processing capabilities of quantum algorithms depends on the quantum implementation of machine learning algorithms involving nonlinear functions operating on information represented topographically, as is common in neural cortex. Several approaches to quantum computation for machine learning by means of topographic representation have been proposed. These approaches show ways to construct unitary operators, implementable on quantum computers, that implement arbitrary (including nonlinear) functions via computational maps.

Optimization problems are ubiquitous in machine learning and other domains like engineering and a number of specialized techniques for efficiently solving specific problems have been developed. A novel quantum optimization algorithm based on Adiabatic Quantum Computation through local evolution and inspired by the ideas from Grover's Search algorithm [1] can be evolved for finding minimum and employed in solving unconstrained optimization problems.

As a transition from the classical paradigm to quantum paradigm, several researchers are involved in developing quantum inspired algorithms to bridge the gap between the two paradigms. Notable among them include the quantum inspired metaheuristic algorithms capable of delivering outputs in real time. Different incarnations of the quantum inspired metaheuristic algorithms have come up, thanks to the relentless efforts on the part of the scientists. These algorithms are efficient enough to obtain stable outputs by emulating the quantum mechanical principles on classical computers. In most of the cases, it is found that the quantum inspired versions of the metaheuristic algorithms outperform their classical counterparts.

Although most of the quantum machine learning algorithms are based on the bilevel qubits, of late, much work is being carried out using multilevel quantum systems (using qutrits or qudits), which promise more efficiency.

https://doi.org/10.1515/9783110670707-006

References

[1] Lavor, C., Manssur, L.R.U., & Portugal, R. Grover's Algorithm: Quantum Database Search. arXiv:quant-ph/0301079.

[2] Beom, H. R., & Cho, H. S. A sensor-based navigation for a mobile robot using fuzzy logic and reinforcement learning. IEEE Transactions on Systems, Man, and Cybernetics. 1995, 25 (3), 464–477.

[3] Connolly, C. I. Harmonic functions and collision probabilities. The International Journal of Robotics Research. 1997,16(4), 497–507.

[4] Smart, W. D., & Kaelbling, L. P. Effective reinforcement learning for mobile robots. Proceedings – IEEE International Conference on Robotics and Automation. 2002, 3404–3410.

[5] Kondo, T., & and Ito, K. A reinforcement learning with evolutionary state recruitment strategy for autonomous mobile robots control. Robotics and Autonomous Systems. 2004, 46(2), 111–124.

[6] Tzafestas, S. G., & Rigatos, G. G. Fuzzy reinforcement learning control for compliance tasks of robotic manipulators. IEEE Transactions on Systems, Man, and Cybernetics. 2002, 32(1), 107–113.

[7] Matsuda, Y., Nishimori, H. & Katzgraber, H. G. Ground-state statistics from annealing algorithms: Quantum vs classical approaches. 2009. arXiv:0808.0365.

[8] Santoro, G. E., Martonak, R., Tosatti, E. & Car, R. Theory of quantum annealing of an Ising spin glass. Science. 2002, 295(5564),2427–2430.

[9] da Silva, A. J., Ludermir, T. B., & de Oliveira, W. R. Quantum perceptron over a field and neural network architecture selection in a quantum computer. Neural Networks. 2016, 76, 55–64.

[10] Panella, M., & Martinelli, G. Neural networks with quantum architecture and quantum learning. International Journal of Circuit Theory and Applications. 2011, 39, 61–77.

[11] Kak, S. On quantum neural computing. Advances in Imaging and Electron Physics. 1995, 94, 259.

[12] Chrisley, R. Quantum Learning. New directions in cognitive science: Proceedings of the international symposium, Saariselka, 4–9 August 1995, Lapland, Finland. Pylkkänen, P. & Pylkkö, P. (Eds.). Finnish Association of Artificial Intelligence, Helsinki, 1995, 77–89.

[13] Narayanan, A., & Menneer, T. Quantum artificial neural network architectures and components. Information Sciences. 2000, 128, 231–255.

[14] Schuld, M., Sinayskiy, I., & Petruccione, F. Simulating a perceptron on a quantum computer. 2014. ArXiv:1412.3635.

[15] Purushothaman, G., & Karayiannis, N. Quantum Neural Networks (QNN's): Inherently Fuzzy Feedforward Neural Networks. IEEE Transactions on Neural Networks. 1997, 8(3), 679.

[16] Behrman, E. C., Steck, J. E., Kumar, P., & Walsh, K. A. Quantum Algorithm design using dynamic learning. Quantum Information and Computation. 2008, 8(1&2), 12–29.

[17] Ventura, D., & Martinez, T. A quantum associative memory based on Grover's algorithm. Proceedings of the International Conference on Artificial Neural Networks and Genetics Algorithms. 1999, 22–27.

Index

https://doi.org/10.1515/9783110670707-007

De Gruyter Frontiers in Computational Intelligence

Already published in the series

Volume 5: Machine Learning Applications
S. Bhattacharyya, R. Das, S. Nandy (Eds.)
ISBN 978-3-11-060853-3, e-ISBN (PDF) 978-3-11-061098-7,
e-ISBN (EPUB) 978-3-11-060866-3

Volume 4: Intelligent Decision Support Systems
S. Bhattacharyya, S. Borra, M. Bouhlel, N. Dey (Eds.)
ISBN 978-3-11-061868-6, e-ISBN (PDF) 978-3-11-062110-5,
e-ISBN (EPUB) 978-3-11-061871-6

Volume 3: Big Data Security
I. Banerjee, S. Bhattacharyya, S. Gupta (Eds.)
ISBN 978-3-11-060588-4, e-ISBN (PDF) 978-3-11-060605-8,
e-ISBN (EPUB) 978-3-11-060596-9

Volume 2: Intelligent Multimedia Data Analysis
S. Bhattacharyya, I. Pan, A. Das, S. Gupta (Eds.)
ISBN 978-3-11-055031-3, e-ISBN (PDF) 978-3-11-055207-2,
e-ISBN (EPUB) 978-3-11-055033-7

Volume 1: Machine Learning for Big Data Analysis
S. Bhattacharyya, H. Baumik, A. Mukherjee, S. De (Eds.)
ISBN 978-3-11-055032-0, e-ISBN (PDF) 978-3-11-055143-3,
e-ISBN (EPUB) 978-3-11-055077-1

Bei Fragen zur Produktsicherheit wenden Sie sich bitte an:
If you have any questions regarding product safety,
please contact:

Walter de Gruyter GmbH
Genthiner Straße 13
10785 Berlin
productsafety@degruyterbrill.com